M000309723

GALLIPOLI

GREAT BATTLES

GALLIPOLI

JENNY MACLEOD

OXFORD
UNIVERSITY PRESS

OXFORD
UNIVERSITY PRESS

Great Clarendon Street, Oxford, OX2 6DP,
United Kingdom

Oxford University Press is a department of the University of Oxford.
It furthers the University's objective of excellence in research, scholarship,
and education by publishing worldwide. Oxford is a registered trade mark of
Oxford University Press in the UK and in certain other countries

First Edition published in 2015

Impression: 1

Published in the United States of America by Oxford University Press
198 Madison Avenue, New York, NY 10016, United States of America

British Library Cataloguing in Publication Data
Data available

Library of Congress Control Number: 2014957912

ISBN 978–0–19–964487–2

Printed in Italy by L.E.G.O. S.p.A.

To Rich
Thanks for the html, css, and the tlc

FOREWORD

For those who practise war in the twenty-first century the idea of a 'great battle' can seem no more than the echo of a remote past. The names on regimental colours or the events commemorated at mess dinners bear little relationship to patrolling in dusty villages or waging 'wars amongst the people'. Contemporary military doctrine downplays the idea of victory, arguing that wars end by negotiation not by the smashing of an enemy army or navy. Indeed it erodes the very division between war and peace, and with it the aspiration to fight a culminating 'great battle'.

And yet to take battle out of war is to redefine war, possibly to the point where some would argue that it ceases to be war. Carl von Clausewitz, who experienced two 'great battles' at first hand—Jena in 1806 and Borodino in 1812—wrote in *On War* that major battle is 'concentrated war', and 'the centre of gravity of the entire campaign'. Clausewitz's remarks related to the theory of strategy. He recognized that in practice armies might avoid battles, but even then the efficacy of their actions relied on the latent threat of fighting. Winston Churchill saw the importance of battles in different terms, not for their place within war but for their impact on historical and national narratives. His forebear, the Duke of Marlborough, commanded in four major battles and named his palace after the most famous of them, Blenheim, fought in 1704. Battles, Churchill wrote in his life of Marlborough, are 'the principal milestones in secular history'. For him, 'Great battles, won or lost, change the entire course of events, create new standards of values, new moods, new atmospheres, in armies and nations, to which all must conform'.

Clausewitz's experience of war was shaped by Napoleon. Like Marlborough, the French emperor sought to bring his enemies to battle. However, each lived within a century of the other, and they fought their wars in the same continent and even on occasion on adjacent ground. Winston Churchill's own experience of war, which spanned the late nineteenth-century colonial conflicts of the British Empire as well as two world wars, became increasingly distanced from the sorts of battle he and Clausewitz described. In 1898 Churchill rode in a cavalry charge in a battle which crushed the Madhist forces of the Sudan in a single day. Four years later the British commander at Omdurman, Lord Kitchener, brought the South African War to a conclusion after a two-year guerrilla conflict in which no climactic battle occurred. Both Churchill and Kitchener served as British Cabinet ministers in the First World War, a conflict in which battles lasted weeks, and even months, and which, despite their scale and duration, did not produce clear-cut outcomes. The 'Battle' of Verdun ran for all but one month of 1916 and that of the Somme for five months. The potentially decisive naval action at Jutland spanned a more traditional twenty-four-hour timetable but was not conclusive and was not replicated during the war. In the Second World War, the major struggle in waters adjacent to Europe, the 'Battle' of the Atlantic, was fought from 1940 to early 1944.

Clausewitz would have called these twentieth-century 'battles' campaigns, or even seen them as wars in their own right. The determination to seek battle and to venerate its effects may therefore be culturally determined, the product of time and place, rather than an inherent attribute of war. The ancient historian Victor Davis Hanson has argued that seeking battle is a 'western way of war' derived from classical Greece. Seemingly supportive of his argument are the writings of Sun Tzu, who flourished in warring states in China between two and five centuries before the birth of Christ, and who pointed out that the most effective way of waging war was to avoid the risks and dangers of actual fighting. Hanson has provoked strong criticism: those who argue that wars can be won without battles are not only to be found in Asia. Eighteenth-century European commanders, deploying armies in close-order formations in order to deliver concentrated fires, realized that the destructive consequences of battle for their own troops could be self-defeating. After the First World War, Basil Liddell Hart developed a theory of strategy which he called 'the indirect approach', and suggested that manoeuvre might substitute for hard fighting, even if its success still relied on the inherent threat of battle.

The winners of battles have been celebrated as heroes, and nations have used their triumphs to establish their founding myths. It is precisely for these reasons that their legacies have outlived their direct political consequences. Commemorated in painting, verse, and music, marked by monumental memorials, and used as the way points for the periodization of history, they have enjoyed cultural afterlives. These are evident in many capitals, in place names and statues, not least in Paris and London. The French tourist who finds himself in a London taxi travelling from Trafalgar Square to Waterloo Station should reflect on his or her own domestic peregrinations from the Rue de Rivoli to the Gare d'Austerlitz. Today's Mongolia venerates the memory of Genghis Khan while Greece and Macedonia scrap over the rights to Alexander the Great.

This series of books on 'great battles' tips its hat to both Clausewitz and Churchill. Each of its volumes situates the battle which it discusses in the context of the war in which it occurred, but each then goes on to discuss its legacy, its historical interpretation and reinterpretation, its place in national memory and commemoration, and its manifestations in art and culture. These are not easy books to write. The victors were more often celebrated than the defeated; the effect of loss on the battlefield could be cultural oblivion. However, that point is not universally true: the British have done more over time to mark their defeats at Gallipoli in 1915 and Dunkirk in 1940 than their conquerors on both occasions. For the history of war to thrive and be productive it needs to embrace the view from 'the other side of the hill', to use the Duke of Wellington's words. The battle the British call Omdurman is for the Sudanese the Battle of Kerreri; the Germans called Waterloo 'la Belle Alliance' and Jutland Skagerrak. Indeed the naming of battles could itself be a sign not only of geographical precision or imprecision (Kerreri is more accurate but as a hill rather than a town is harder to find on a small-scale map), but also of cultural choice. In 1914 the German general staff opted to name their defeat of the Russians in East Prussia not Allenstein (as geography suggested) but Tannenberg, in order to claim revenge for the defeat of the Teutonic Knights in 1410.

Military history, more than many other forms of history, is bound up with national stories. All too frequently it fails to be comparative, to recognize that war is a 'clash of wills' (to quote Clausewitz once more), and so omits to address both parties to the fight. Cultural difference and, even more, linguistic ignorance can prevent the historian considering a battle in the round; so too can the availability of sources. Levels of literacy matter here, but so does cultural survival. Often these pressures can be congruent but they can also be divergent. Britain enjoys much higher levels of literacy than Afghanistan, but in 2002 the memory of the two countries' three wars flourished in the latter, thanks to an oral tradition, much more robustly than in the former, for whom literacy had created distance. And the historian who addresses cultural legacy is likely to face a much more challenging task the further in the past the battle occurred. The opportunity for invention and reinvention is simply greater the longer the lapse of time since the key event.

All historians of war must, nonetheless, never forget that, however rich and splendid the cultural legacy of a great battle, it was won and lost by fighting, by killing and being killed. The Battle of Waterloo has left as abundant a footprint as any, but the general who harvested most of its glory reflected on it in terms which have general applicability, and carry across time in their capacity to capture a universal truth. Wellington wrote to Lady Shelley in its immediate aftermath: 'I hope to God I have fought my last battle. It is a bad thing to be always fighting. While in the thick of it I am much too occupied to feel anything; but it is wretched just after. It is quite impossible to think of glory. Both mind and feelings are exhausted. I am wretched even at the moment of victory, and I always say that, next to a battle lost, the greatest misery is a battle gained.' Readers of this series should never forget the immediate suffering caused by battle, as well as the courage required to engage in it: the physical courage of the soldier, sailor, or warrior, and the moral courage of the commander, ready to hazard all on its uncertain outcomes.

HEW STRACHAN

PREFACE

This book presents a military narrative of the Gallipoli campaign and its memory during the subsequent century which encompasses five key participant nations: Australia, New Zealand, Great Britain, Ireland, and the Ottoman Empire/Turkey. This has been made possible by two important developments. The first is the increased use of Ottoman military sources by historians. This has resulted in fundamental changes in the assessments made of the campaign. The earlier neglect of the Ottoman 'side of the hill' was due to the inaccessibility of the Turkish sources because of bureaucratic barriers and the language and script in which they are written. Furthermore, the low rates of literacy amongst the rank and file (below 10 per cent in the Ottoman Army)[1] have traditionally been cited as a cause of the very limited numbers of private records of the campaign. Thanks to the work of Ed Erickson, Tim Travers, Robin Prior, Mesut Uyar, and others, it is now becoming possible to write the Ottoman role back into the Gallipoli story.[2] The military section of this book stands on their shoulders. It will argue that the Allied attempt to invade the Ottoman Empire was a flawed strategy whose execution was botched by the nature of its inception, but also that the skill and determination of the Ottoman forces along with their superior tactical positions ensured Ottoman victory.

The second change is that technology, and specifically the digitization of sources, have made it possible to research this book in a quite different way from my PhD days in the 1990s.[3] Access to broadly comparable bodies of digitized newspapers from five nations has enabled both breadth of study and a longitudinal comparison of changing perceptions.[4] This has revealed the way in which ideas and attitudes have reverberated across increasingly disparate societies over time.

This book, and the series to which it contributes, takes the novel approach of presenting an extended essay on the military campaign before analysing its memory and commemoration at length. It does so in order to highlight the relationship between history and memory, and the particularly selective and flexible nature of the latter. Memory, sometimes referred to as collective memory or myth, is the subjective way in which a group remembers an event. Memory, it has been said, 'accommodates only those facts that suit it'.[5] It will be seen that the memory of the campaign has varied a great deal.

Thus, for example, if we take history books as the prism through which we study perceptions of the campaign, we can see that the author inevitably reveals the priorities and attitudes of the day and their location, however much he or she may strive for objectivity. In the case of Gallipoli, it has been portrayed at different times, for example, as symbolizing the unity of empire or, conversely, as exemplifying the Australian character. The same is true of commemoration, which to an important extent is always present-minded.[6] That is to say, the commemoration of war is, at its core, a heartfelt tribute to the participation and deaths of others. But it also reflects what those doing the commemorating thought that meant and what is important to society at that moment. We can explore what the participation and deaths of people in war was thought to mean by examining the arrangements for the commemoration, and through the speeches, sermons, and editorials written for the occasion. Similarly, the design, finance, and construction of war memorials can be revealing. It was forty-five years before Turkey chose to add to the modest Sergeant Mehmet Memorial (Mehmed Çavuş Anıtı) on the peninsula by building the towering Çanakkale Martyrs' Memorial—why didn't it seem important to do so before then? We can further explore the significance attached to the campaign by tracing attendance at commemorative events, the media coverage it attracted, and through related events such as book reviews, films on the subject, or pilgrimages to special sites of memory on key anniversaries.

Given the scope of the book, a range of challenges have arisen. The vast complexity of the movements of thousands of men has had to be simplified, and a broad overview featuring key incidents offered in its place. It is often difficult to marry up events described from opposing perspectives. The use of different names for locations on the peninsula has not helped. Where possible, I have tried to acknowledge all of the names used at the earliest convenient point, and then I have reverted to the most familiar, even though this sometimes amounts to a kind of nominal imperialism. Despite it being more common to describe what happened on 25 April 1915 as 'landings', in trying to present matters from the Ottoman point of view as far as possible, I have sometimes referred to events as an invasion.[7] Finally, in the light of the diversity within the Ottoman Empire and its Fifth Army, it is very important to refer to its members as Ottomans, and not Turks as many studies do. To do the latter is to participate in the Republic of Turkey's subsequent attempts to erase much of its multi-ethnic past. This book attempts to be consistent in referring to 'Ottomans' and the 'Ottoman Empire' when referring to the period pre-1923, and to 'Turks' and 'Turkey' only thereafter.

JENNY MACLEOD
February 2015

ACKNOWLEDGEMENTS

The idea of writing this book came from the series editor Hew Strachan, and I am indebted to him for the opportunity he has given me, and for his close support in the editing process. His invitation letter was waiting in my pigeon hole when I returned to work from my first maternity leave. I hadn't written about Gallipoli in about ten years and was in need of a new project. In the intervening decade, the way I think about history has changed considerably, and much of that is down to what I've learned from the International Society for First World War Studies. This most collegial and knowledgeable group of scholars, and the society's president, Pierre Purseigle (who it must be noted, is also an excellent part-time choreographer), have influenced the book in all sorts of ways. Among those who helped with queries through the society were Hedley Malloch, Jason Engle, James Kitchen, Peter Stanley, and Giorgio Rota.

Early discussions with Carl Bridge, Stuart Ward, and Jeff Grey shaped this book. Stuart put me in touch with Asiye Sanli Koca who helped me to develop my initial thoughts about the Turkish memory of the campaign. Thereafter Altay Atli taught me an extraordinary amount about Turkish history, and found some wonderful sources for me. He then put me in touch with Gizem Tongo, who has worked more formally as my research assistant. Her superb research led to her drafting an early version of Chapter 8 in answer to my questions, and indeed, without Gizem and Altay the most original part of this book would not have been possible. I am also grateful to Yan Overfield Shaw who assisted in translating the

Turkish sources into English. Mustafa Onur Yurdal has kindly allowed me to use some rare photos of Anzac Cove. In addition, Tom Burke, Camilla Russell, Aaron Culbertson, and Brian Hughes found useful material for me, and stretching back to pre-digital days and ever since, Zoë Greenwood, Adam Barclay, and Mary and Oliver Howie have kept watch for interesting items in the Australian press for me. Michael Macaulay very thoughtfully emigrated to New Zealand just as I needed someone to track down some additional material in Wellington, and has been incredibly generous with his time and research expertise. The same generosity has been marked by numerous colleagues who have read and commented on various draft chapters. Profound thanks are due to: Altay Atli, Ayhan Aktar, James Bennett, Ed Erickson, Ian McGibbon, Gencer Özcan, Catriona Pennell, Jock Phillips, Mesut Uyar, and especially to John Connor who shared his press cuttings with me and also tracked down key information in the National Library of Australia. Dennis Showalter read the earliest scrappiest draft, and Helen McCartney and Jay Winter have read it all from start to finish. Harvey Broadbent, Rhys Crawley, Richard Grayson, Keith Jeffery, John Lee, Henry Schydlo, Robin Prior, and Brad West have been among those who have answered my specific queries. The University of Hull gave me a semester's research leave, and my colleagues there have been kind and supportive. Thanks are particularly due to Mandy Capern, Nick Evans, and Doug Hamilton, and to David Starkey and Peter Wilson. At OUP, Matthew Cotton has been patience personified and Erica Martin has done a great job locating suitable pictures. Thanks also to Luciana O'Flaherty and all the other OUP staff behind the scenes. All of the book's shortcomings, of course, are entirely mine.

Finally, to all my friends and family who have encouraged me along the way or tolerated my absence as I've batted onwards through the research and writing process, thank you one and all. My lovely little girls, Molly and Sophie, are a great joy to me and have been very understanding when their Mum disappeared off to work when I should have been playing with them. Any working parent knows that a good backup team is vital. My mother-in-law, Patty Halford and sister-in-law, Jane Halford, have been superlative in this regard. My Mum, Jacky Macleod, practically moved in to enable me to finish this thing. And the last word must go to my clever and lovely husband Rich, for his unstinting support, love, and enthusiasm.

CONTENTS

LIST OF FIGURES

LIST OF MAPS

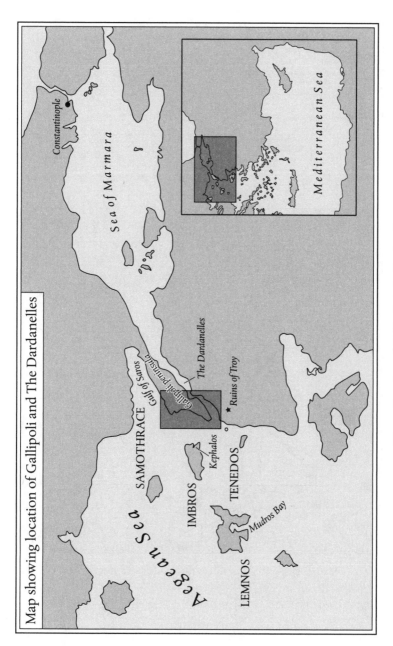

Map 1. Location of Gallipoli at the Dardanelles

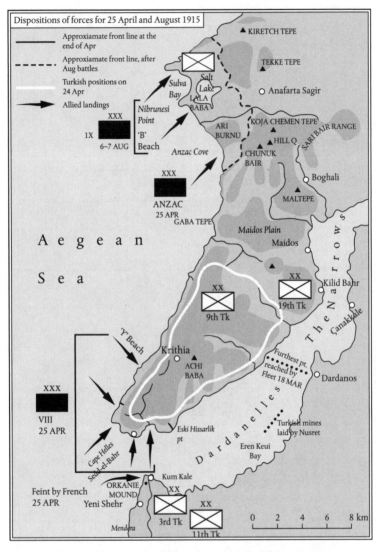

Map 2. Dispositions of forces for 25 April and August 1915

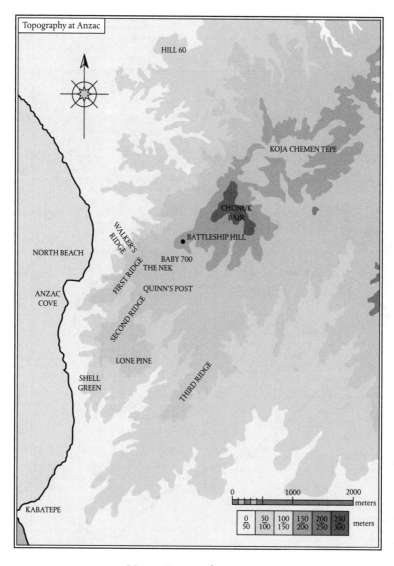

Map 3. Topography at Anzac

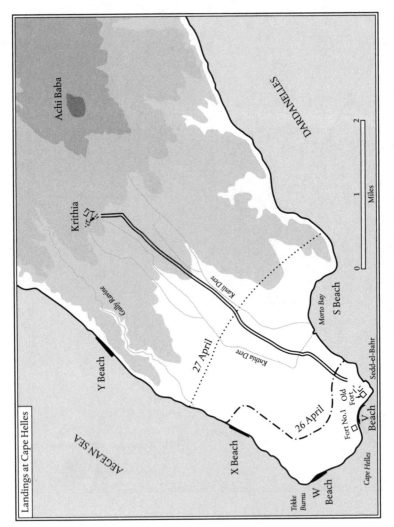

Map 4. Landings at Cape Helles

1

Introduction

In the early hours of 19 May 1915, 50,000 soldiers from the Ottoman Empire attacked the Australian and New Zealand soldiers who had invaded their territory. These 'Anzacs' (from the acronym for the Australian and New Zealand Army Corps) had fought ferociously to gain a small area of the Gallipoli peninsula since dramatic amphibious landings on 25 April. Now the Ottoman Northern Group,[1] its morale high and freshly reinforced, aimed to drive them into the sea. But the Ottomans lacked both adequate artillery backup and the element of surprise. A reconnaissance flight over the peninsula by the Royal Naval Air Service had spotted the disembarkation of large numbers of Ottoman reinforcements at Akbas pier. For their part, the Anzacs had noticed a let-up in activity on 18 May, suggesting that preparations were under way. That night they were ready for an attack. After a preliminary bombardment, wave after wave of Ottoman soldiers came forward determinedly. The men in 2nd Division steadily called out 'Allah Allah' and sang the *Motherland March*. All of them were met by murderous machine-gun and rifle fire. By 11.20 a.m. the following day, III Corps commander Brigadier General Esat reported the attack had failed.[2] The defence of their lines cost the Anzacs 628 casualties,[3] while the Ottomans suffered over 10,000 casualties, of whom about 3,500 died. The commander of the Ottoman forces, General Otto Liman von Sanders, later admitted, 'I feel that the attack was an error on my part based on an underestimation of the enemy.'[4]

The bodies lay thickly piled in no man's land. In some sectors, such as by Quinn's Post, the area was so narrow that sometimes the enemy

1

could be heard talking in the opposing trenches. Now, in the warm weather of early summer, the mass of corpses began to decompose. The stench was unbearable—it was said that with an offshore breeze, it could be smelled from the boats just off the coast.[5] Negotiations for an armistice began, led by the Turcophile Aubrey Herbert, Conservative MP for South Somerset and an intelligence officer with the New Zealand Division.[6] It took place on 24 May, with men from each side emerging cautiously at first from the shelter of their trenches, before getting to work on burying the swollen and grotesquely decayed bodies in shallow holes in the ground. One witness, Compton Mackenzie, later wrote,

> Looking down I saw squelching up from the ground on either side of my boot like a rotten mangold the deliquescent green and black flesh of a Turk's head. [. . .] I cannot recall a single incident on the way back down the valley. I only know that nothing could cleanse the smell of death from the nostrils for a fortnight afterwards.[7]

Herbert spent the day in the company of an Ottoman liaison officer, who told him, 'At this spectacle even the most gentle must feel savage, and the most savage must weep.'[8] Little wonder, then, that after this experience the mood at Anzac changed considerably. Hatred was replaced with something more like respect, the lurid tales of the Ottomans' savagery and brutality faded away and turned to sympathy for 'Johnny Turk', a fellow sufferer. This is one of the most famous incidents in the ten-month-long attempt by the forces of the British Empire, in concert with her French ally, to knock the Ottoman Empire out of the war and to find an alternative to the stalemate of the Western Front. It failed on both counts.

The situation following the 19 May attack was but the most acute example of a broader problem on the peninsula and at Anzac in particular. The area that had been captured was relatively small, the terrain difficult: steep, arid, and jagged. Contrary to expectations, there was no expanse of rear area in which to safely situate supplies, medical facilities, or places to relax. They had their backs to the beach, and nowhere was safe from Ottoman gunfire: even swimming in the sea, the one respite men had from the tension or the summer's heat, was fraught with danger. Nor was it a simple matter to set up appropriate latrine facilities. A plague of flies buzzed between the dead bodies, the latrines, and the men's food. The consequent toll of sickness from dysentery or typhoid fever steadily mounted. A chronic shortage of water and of the means to sterilize what little water they had, made daily life difficult, and diarrhoea deadly.[9] The evacuation of the sick and wounded in these circumstances was extremely difficult. It was particularly difficult at Anzac to gather the wounded from the front lines and to evacuate them. Most famously, an ambulance man from South Shields in the north of England attached to the Australian Imperial Force (AIF), John Simpson Kirkpatrick, found an ad hoc solution to the problem and commandeered a donkey to transport lightly wounded men down the valleys away from the front line. He died on 19 May.

Simpson and his donkey have since achieved iconic status in Australia. For the events at Gallipoli became the basis of a foundational myth for Australia, known as the Anzac legend. This is a description of the qualities ascribed to the Australian soldiers who fought in the Gallipoli campaign, which has been projected on to all subsequent Australian soldiers, and beyond them, on to the nation itself. The description has not remained static: it has evolved over time. Simpson, for example, was only celebrated from the 1960s, including on the commemorative stamp for the fiftieth anniversary of the Gallipoli campaign. Yet, by 1988 he was the subject of a life-size statue prominently positioned at the Australian War Memorial.[10] Long before Simpson's elevation, however, the Anzacs were assiduously remembered as heroes: in newspaper reports and editorials, recruiting speeches and church sermons, propaganda and history books, and above all in commemorations of the anniversary of the Gallipoli landings on 'Anzac Day'.

On the centenary of the first Anzac Day on 25 April 2015, 10,500 Australians and New Zealanders will gather at Anzac Cove on the

Gallipoli peninsula in Turkey to commemorate the landing under fire by Australian soldiers. Such was the anticipated interest in attending this event that the Australian and New Zealand governments held a ballot to allocate 8,000 and 2,000 places respectively, with 500 places reserved for official representatives of all nations involved in the campaign. The Australian part of the ballot was more than five times oversubscribed.[11]

The existence of the Anzac Cove ballot and its division of places on a ratio of 8:2:0.5 (Australia:New Zealand:other nations), reflects the limited space at Anzac Cove and the number of casualties suffered by Australia and New Zealand in relation to each other during the campaign.[12] The last part of the ratio in no way reflects the casualties suffered by the other nations who fought at Gallipoli. But it does indicate something about the different levels of interest in the campaign elsewhere. No such ballot was needed for places at the Turkish, French, or Commonwealth services the day before, even though they represent the commemoration of far more populous countries, far larger armies involved in the campaign, and far more numerous dead and wounded soldiers.

This disparity is because, through the Anzac legend, the Gallipoli campaign has achieved an importance in Australia and, to a slightly lesser extent to New Zealand, which goes far beyond the sacrifice of loss in war. The legend's importance is reflected in the way that Anzac Day has become these countries' de facto national day. It overshadows Australia Day and Waitangi Day (and their difficult messages about the usurpation of indigenous peoples by British settlement). The Anzac legend enabled Australia and New Zealand to forge new national identities for themselves which overwrote their pre-existing Britishness. It also served to stunt more radical interpretations of how these Antipodean nations saw themselves. Before Anzac, Australia saw itself as a workers' paradise, while New Zealand was thought of as a social laboratory that led the world in granting women the franchise.[13]

Of course, they were not the only countries forging their national identities in the years following the First World War. Perhaps the most spectacular territorial consequence of the war was the breakup of the Ottoman Empire. The Republic of Turkey, established in 1923, was the largest of its successor states. Its founding president, Mustafa Kemal, played what is now seen as a pivotal role in the Ottoman victory at Gallipoli, but it was the War of Independence that was most important as a founding myth for the new republic. In the process Turkey set about creating itself as a modern state, and did much to erase the memory of the multi-ethnic Ottoman past with its defeats and massacres.[14] There are similarities here with Ireland, which achieved independence in 1922 after a war of independence with Britain. In consequence, there was limited appetite thereafter to commemorate the role of Irish soldiers in the British Army during 1914–18. It was the heroes of the revolution who were to be remembered in forging a new state.

If it was the perceived Britishness of the war that caused the Irish to forget the First World War for so long, conversely it was a continuing assiduous endeavour to assert their Britishness that, in part, caused the Ulstermen of the newly formed Northern Ireland to commemorate their own role in the First World War, particularly that of the 36th (Ulster) Division on the first day of the Somme.[15] This eclipsed the involvement of Ulstermen in all other campaigns, including Gallipoli. Shorn of its sectarian and unionist aspects, the Somme also remains the key site of memory for the war in the rest of Britain. Thus, although Gallipoli was the most important campaign until July 1916 for Britain, it is less well remembered there. This is, of course, related, to the unparalleled losses of the first day of the five-month-long campaign on the Somme. But it may also be because the British seem to have purposefully relinquished the Britishness of the campaign, to allow it to be a moment of particular pride for Australia and New Zealand from the first Anzac Day in 1916 onwards.

But, as we've seen with Simpson and the donkey in Australia, the remembrance of Gallipoli has not remained fixed and unchanging in any of the participant nations. From the 1960s to the 1980s, the relevance of the Anzac legend seemed to be dying away in Australia and New Zealand; meanwhile the First World War (and therefore Gallipoli too) was scarcely mentioned in Ireland, its national memorial to 1914–18 left to become semi-derelict. But in the southern hemisphere, the renewal of interest and remembrance of the 1990s has built into an unassailable and unavoidable nationwide commemoration of the centenary years.[16] The Australian government plans to spend at least AUD 140 million on centenary commemorations (compared to Great Britain's £55 million (AUD 94 million), and New Zealand's NZD 19 million (AUD 18 million)).[17] Meanwhile, in Ireland, political reconciliation at the turn of the twenty-first century brought the opportunity for depoliticized commemoration.[18] The Irish National War Memorial has been refurbished, a reconciliatory visit by the Queen has taken place, and Gallipoli and the First World War find their place amidst Ireland's official 'Decade of Centenaries'.[19]

Thus, the huge numbers of pilgrims expected at Anzac Cove on 25 April 2015 contrast with the limited few who were able to bear the expense of attending the fiftieth anniversary commemorations in an age before affordable international travel. Interestingly, it was not just Australian, New Zealand, British, and French veterans who went to Anzac Cove and Cape Helles, Gallipoli in 1965, but also German veterans who travelled to commemorate the campaign. However, the two groups did not meet each other—they were not there at the same time or the same place. Whereas Britain, Australia, and New Zealand mark 25 April as the pivotal date in the Gallipoli campaign, Turkey commemorates their naval victory of 18 March as well as 10 August, the anniversary of their victory in the 'Battle of Anafartalar', that is, at Suvla. Thus, for the fiftieth anniversary commemorations, German (and also a group of Irish) veterans visited Çanakkale in March whilst 285 Anzacs (including women who had served as nurses), plus French and 'English' veterans (as they were termed by the Turkish newspaper reporting the event), attended in April.[20] These groups had found quite different moments of the campaign to be significant.

This book is a study of the Gallipoli campaign and its remembrance. It aims to present the most fully transnational examination of the campaign and its memory that has been written to date. That is, the familiar Australian aspects of the campaign and its memory, extraordinary and important in their own right, are studied here in comparison to the participation of other nations.[21] For despite the intrinsic importance of the actions of the AIF for Australia, it must be admitted that the AIF played a supporting role in the Gallipoli campaign, at least until the doomed August offensive that it spearheaded from Anzac Cove. Rather, the attack on Gallipoli was a joint and combined effort by forces from the British and French empires. As such, men from England, Scotland, Ireland (north and south), and Wales fought alongside men from Australia and New Zealand, including a Maori Detachment, as well as Newfoundlanders, a sizeable contingent from India and Ceylon, and the 500-strong Zion Mule Corps, reputed to be 'the first all-Jewish military force to be in action for more than two thousand years'.[22] The French Corps Expéditionnaire d'Orient (CEO) comprised recruits from metropolitan France, French West Africa, and the Foreign Legion. The Ottoman forces largely comprised ethnic Turks from Anatolia, but also an Arab regiment, and Kurdish, Greek, Armenian, and Jewish soldiers as well.[23] The most senior officers in the Ottoman Fifth Army, including its commander, Liman von Sanders, were German, whilst Austria-Hungary provided several artillery units.

The following chapters aim to examine the campaign from both sides of the hill, and then to study and compare how it has been remembered in five countries, and ultimately, to show how they have affected one another of late. It aims thereby to contribute to our collective understanding of the First World War as a global war that has had a remarkably deep and long-lasting impact upon many nations.[24]

2

Origins

The Ottoman Empire came under attack from British and French warships on 19 February 1915 at the mouth of the Dardanelles straits, which guard the route from the Mediterranean Sea through to the Sea of Marmara and on to the imperial capital, Constantinople. After a series of attacks, it saw off the naval onslaught a month later on 18 March, only for the Allied forces to regroup and renew their assault through a series of amphibious landings on 25 April. On that day, the Mediterranean Expeditionary Force (MEF) gained some precarious toeholds on the difficult terrain of the Gallipoli peninsula which stands on one side of the Dardanelles. Despite some tenacious fighting, and a further concerted assault on the peninsula in August, the armed forces of the Ottoman Empire defeated the invasion, and the MEF evacuated their positions in two stages in December 1915 and January 1916.

This was a rare victory for the Ottoman Empire. In its 600-year existence, it had at times incorporated vast swathes of the Near East, reaching from the Arabian peninsula in the south, westwards to Algeria in northern Africa, east towards Persia, and northwards into Europe, famously reaching as far as the gates of Vienna in 1683. Yet from the late nineteenth century in particular, the empire's decline was apparent, and the Great Powers hovered, ready to pick away at its lands and encroach on its day-to-day business. The loss of control over significant parts of its territory ensued.[1] A revolution in 1908 led by young army officers, the Committee of Union and Progress (CUP), better known to posterity by the misnomer 'the Young Turks', attempted to bring stronger government but had limited overall impact in stemming the tide of the empire's decline. After war in Tripolitania (modern-day Libya) and humiliation in the Balkans between 1908 and 1913, the Ottomans chose to enter the First World War on the German side on 29 October 1914. It proved to be a disastrous decision, and their part in the Central Powers' overall defeat in the war brought with it the end of the Ottoman Empire.[2]

The choice of Germany as its ally, sealed in a secret treaty signed on 2 August 1914, had seemed to hold the best prospects for the Ottoman Empire of preventing further territorial loss and foreign domination. Russia seemed particularly menacing, and her ally Britain showed minimal interest in securing Ottoman support as the war clouds gathered. Nonetheless, the Ottoman Empire remained neutral for some months, hoping for a short war that would not require its active intervention. Until any declaration of war, the theoretical possibility remained that it could join either side. Britain and Germany were both closely involved in its armed forces.[3] Britain had had a naval mission in Constantinople since 1908, and from December 1913 General Liman von Sanders led a German military mission charged with modernizing the Ottoman Army.[4] Yet the writing was on the wall as early as the middle of August, when Britain's Rear Admiral Limpus and his men were called back from their ships to desk jobs in the Ottoman naval ministry.[5]

It is tempting to see the Ottoman decision as a tale of four warships, a tale which exemplifies the attitudes of the two potential allies. Two battleships were under construction in Britain for the Ottoman government: the *Reşadiye* was being built in Barrow-in-Furness, the *Sultan Osman I* on the Tyne. They had been paid for by public subscription and thus great emotion was attached to them. In the summer of 1914, they were nearly complete. But while the Ottoman crew was waiting on Tyneside to take up their posts on their new battleship, Winston Churchill, Britain's First Lord of the Admiralty, ordered that firms building foreign ships should not permit them to pass into foreign hands. He had summarily requisitioned them.[6] The righteous outrage

Fig. 1. General Otto Liman von Sanders.

felt in the Ottoman Empire was stoked and exploited by Germany who sent their Mediterranean squadron to the Dardanelles, narrowly evading the British and French fleets hunting for them. In an audacious move, the *Goeben* and the *Breslau* were sold for a nominal sum to the Ottoman Empire, hoisted the Ottoman flag, and passed through the Dardanelles straits on 10 August. They were subsequently renamed the *Yavuz Sultan Selim* and *Midilli*, but they continued to be commanded by Admiral Souchon and his German crew (now each wearing a fez).[7] This effectively sealed Germany's diplomatic dominance of the Ottoman Empire. Yet there were several weeks of intense pressure from Berlin before the Ottoman Empire heralded its declaration of war against the entente with Admiral Souchon's pre-emptive naval attack on the Russian fleet in the Black Sea on 29 October 1914. Russia, France, and Britain reciprocated with declarations of war in early November.

The Decision to Attack at the Dardanelles

Even before the Ottoman Empire became a belligerent, an attack on the Dardanelles straits was under discussion in London, as the bewildered leaders of the British Empire sought a quick and imaginative resolution to the war.[8] Until its outbreak, the British Army was geared up to fight small colonial wars, and had not fought in a large Continental war for a century. The defence of the realm rested instead on the Royal Navy, and when in the early 1900s a threat from Kaiser Wilhelm's Germany had seemed to emerge there had been a clamour for more ships not more soldiers. Yet in 1914, British traditions were turned on their head as a new and shocking type of warfare was revealed. The new secretary of state for war, Kitchener, set about raising a citizens' army to deploy on the Continent, and the Royal Navy found itself in a supporting role. This was an intolerable situation for the ambitious and imaginative First Lord of the Admiralty, Winston Churchill, who set about finding a leading role for the forces in his command.

The possibility of an attack on the Dardanelles was mooted as early in the war as August 1914 by Sir Louis Mallet, British ambassador to the Ottoman Empire. Its first guise of a Greek-led attack at the behest of Russia soon melted away.[9] Nonetheless, the idea led General Charles Callwell, director of military operations, to examine the prospect for Churchill. Callwell's memorandum of 3 September concluded that a 60,000-strong army would be required to capture the Gallipoli peninsula to enable the fleet to pass through the Dardanelles.[10] The operation, he concluded, would be 'extremely difficult'. This was consistent with all previous appraisals of such an attack. In 1906, a report by the General Staff at the War Office had counselled against attacking the Ottoman Empire via the Dardanelles, and since then significant work had been undertaken to strengthen the area's defences. On 3 November, two days before the British declaration of war on the Ottoman Empire, the British squadron at the mouth of the Dardanelles had bombarded the forts on either side at Sedd-el-Bahr and Kum Kale for ten minutes, under orders from Churchill. A lucky hit on the magazine at Sedd-el-Bahr had caused a spectacular explosion that destroyed its heavy guns and killed eighty-six men.[11]

The possibility of attacks on Gallipoli and the Dardanelles were discussed at the War Councils of 25 and 30 November, but when it became clear that Greek forces would not be forthcoming for an attack, the idea fell into abeyance until the end of the year. The situation altered from 2 January 1915, when a telegram was received from Russia's Grand Duke Nicholas. Facing strong attacks from Ottoman forces at the Battle of Sarikamiş in the Caucasus Mountains, Russia requested that a diversionary attack on the Ottoman Empire should be made to draw off some of the pressure. Britain's promise to do so brought an attack on the Dardanelles back on to the table. Yet while Kitchener initially envisaged a 'demonstration',[12] the excitable Fisher called privately for a joint military and naval attack on the Ottoman Empire: 'I CONSIDER THE ATTACK ON TURKEY HOLDS THE FIELD!—but ONLY if it's IMMEDIATE!'[13] (His idea was utterly impractical, involving two neutral countries and the use of a significant number of troops already engaged in France. It was ignored.) Nonetheless, Churchill sought an alternative means to pursue an attack on the Dardanelles. He contacted Vice Admiral Carden, commander of the British squadron in the Aegean, and canvassed his views on an attack 'by ships alone'. In a consciously manipulative phrase, his telegram to Carden concluded, 'Importance of results would justify severe loss. Let me know your views.'[14]

Carden was initially cautious in his response: 'I do not consider Dardanelles can be rushed. They might be forced by extended operations with large number of ships.'[15] Churchill then encouraged him to flesh out this latter suggestion, and the vice admiral proposed a four-stage attack in response.[16] Churchill won the support in principle of the War Council for this plan on 13 January, and then notified Carden that it would probably go ahead by 15 February when the *Queen Elizabeth* should have arrived in theatre.[17] The idea was a purely naval attack, gradually making its way through the Dardanelles, which would then meet and defeat the Ottoman/German fleet in the Sea of Marmara,[18] and thereafter overawe the Ottoman government. By the War Council meeting of 28 January, the plan agreed in principle was being discussed as an operation that would certainly happen, but one which could be broken off if no satisfactory progress was made.[19] The first naval attack was scheduled for 19 February.

The Making of British Strategy

In the early weeks of 1915, over the course of a series of War Council meetings, the decision to open a second significant theatre of operations for British forces had thus emerged. It was not a decision that was based on a proper assessment of available resources. Nor was it designed in such a way as to ensure the best possible chance of success—that would have been a surprise, coordinated, joint military and naval attack, using well-trained and well-resourced forces. Instead, an attack by ships alone was put into action, despite the grave misgivings of the senior naval advisers in the Admiralty, who knew well enough that naval guns were ill-suited to attacking forts and that the task could not be completed without land forces.

How had this come to pass? The answer lies in a confusing mixture of structures, personalities, cultures, and strategic imperatives. The British government's decision-making apparatus was quite unsuited to conditions of modern warfare and remained so until at least December 1916 and Lloyd George's accession to power. It was four months into the war before the twenty-two-man Cabinet was replaced by a more streamlined War Council for the purpose of directing the war. It still had at least twelve members, it met irregularly, and it was dysfunctional. The prime minister, Herbert Henry Asquith, aged 62, had been in office since 1906 at the head of what had been a remarkably radical government. His style of leadership, however, was not well suited to war: conciliatory, reticent, and cautious, his main interventions appear to have been to draw together the conclusions of a discussion rather than to lead the debate to decision. If the War Council's consideration of strategic options was less than forensic, the blame must lie with Asquith. He nonetheless made some shrewd appointments, and he had some monstrous personalities to deal with. Making Field Marshal Lord Horatio Herbert Kitchener, hero of colonial warfare, secretary of state for war, was a masterstroke in terms of inspiring confidence and enthusiasm in the general public at the outbreak of the war. But Kitchener, a physically imposing man of few words, was ill-suited to the cut and thrust of political debate. Furthermore, he suspected his colleagues could not keep secrets—not unreasonably, since Asquith, for one, was writing very fulsome accounts of government business to Venetia Stanley. Kitchener was unwilling, and perhaps unable, to make his new politician colleagues understand the limits of Britain's capabilities at this stage in the war.[20] Churchill, the ambitious, bumptious, and bold 40-year-old in charge of the Admiralty, ran rings around him, and around his elderly and erratic First Sea Lord Admiral Sir John (Jacky) Fisher. Churchill had recalled the 73-year-old pioneering admiral to office at the outbreak of the war, but Fisher struggled to maintain a consistent viewpoint or to resist Churchill's unswerving pursuit of the Dardanelles idea. Whereas on 3 January, Fisher had been violently in favour of attacking the Dardanelles, by the end of the month he was writing to Jellicoe, 'I just abominate the Dardanelles operation, unless a great change is made and it is settled to be a military operation, with 200,000 men in conjunction with the Fleet.'[21] The next week, on 28 January, Fisher very nearly walked out of a Cabinet meeting, such were his misgivings about the ships-alone plan, but later in the day he accepted the decision and supported it 'totus porcus [the whole hog]'.[22]

Underpinning the assumption that a naval attack would be sufficient to 'take the Gallipoli peninsula, with Constantinople as its objective',[23] was an underestimation of Ottoman resilience. Informed by its recent military defeats and perhaps by the initially pliant response to the late 1912 incursion of a fleet from the Great Powers in the Ottoman capital,[24] the British expected that the fall of the forts would be followed by revolution.[25] But why was an attack on the Ottoman Empire deemed necessary or desirable? The strategic imperatives behind the decision went beyond the desire to bring the Royal Navy into play, and the Grand Duke's appeal for assistance—in any case the emergency in the Caucasus Mountains had passed off with a Russian victory of sorts. The idea of supplying armaments to Russia (and to receive food supplies in return) through a warm water port remained one persuasive factor: not that Britain had any noticeable surplus of armaments to share, nor Russia food. Another factor was the need to defend Egypt, and with it the strategically vital Suez Canal and the route to India. By drawing Ottoman forces to the Dardanelles, any threat to Egypt would be reduced in consequence. This vein of thought may have had some merit. The final strategic imperative, though, did not: to threaten Germany and seek victory in the war overall, by attacking her weaker partner, the Ottoman Empire, and thereby inspire a range of other powers in the Balkans to join the Allied cause.[26] Even Churchill, the Dardanelles scheme's most ardent promoter, expressed doubts about this justification: he told Fisher in a secret memo, 'Germany is the foe, & it is bad war to seek cheaper victories & easier antagonists.'[27]

From Ships Alone to an Invasion Force

The War Council decision of 13 January triggered the Admiralty to gather a substantial fleet of British and French ships for the attack. Carden's squadron had included two French ships since late September, and on 31 January Churchill heard from Victor Augagneur, minister of marine, that four French ships would participate in the attack.[28] Meanwhile discussions in London continued and the War Council drifted towards a far more significant commitment. By 16 February the War Council had decided to send the last remaining British regular division, the 29th, as well as the Australian and New Zealand troops currently in Egypt and Churchill's newly formed Royal Naval Division, to the isle of Lemnos—50,000 men altogether—to be on hand to assist the naval attack.[29] The vague idea seems to have been that they would be there to occupy both Gallipoli and then Constantinople in anticipation of Ottoman resistance rapidly crumbling in the face of the naval assault. Thus the naval attack got under way without them, and indeed on its first day, Kitchener staged a volte-face and decided to keep the 29th Division at home.[30] Even so, the mood in London was changing. On 24 February, Churchill said of the British position at the War Council, 'we were now absolutely committed to seeing through the attack on the Dardanelles'.[31] Amid growing optimism that the attack would succeed, no one appeared to notice this fundamental change in position—no longer was this a demonstration that could be broken off at any point. Furthermore, as historian Robin Prior has pointed out, since military assistance for this naval attack was now on the cards, this meant that a potentially very significant commitment of resources had been made with scarcely any proper formal discussion.[32] On the same day, the French had decided to send 400 officers and 18,000 men to assist in the operations. The main motivation for their involvement was political and not military—if a significant attack was to take place in the Near East, France had to be involved to maintain its national pride and its national interests in the Levant.[33] By early March, with limited progress in theatre, confidence was on the wane, and the men on the spot, Vice Admiral Carden and his second-in-command Vice Admiral De Robeck, had both concluded that military cooperation was required.[34] On 10 March, Kitchener relented and announced that he would release the 29th Division for the Dardanelles.[35] Two days later, he appointed General Sir Ian Hamilton to command the MEF. It is indicative of the extraordinary mission creep and half-baked planning in Whitehall, that the military commander of the Dardanelles expedition was appointed twenty-one days after the naval assault got under way. The operations that began on 19 February were now certain to be the first salvo in a major campaign.

Fig. 2. Commodore Roger Keyes, Vice Admiral John De Robeck, General Sir Ian Hamilton and Major-General Walter Braithwaite on board HMS *Triad* in May 1915.

The Naval Attack

The Dardanelles guard the route to Constantinople and provide the only maritime access to the Black Sea, and hence the only route to Europe for Russian exports that was accessible year-round. The strategic importance of them was such—they have been described as the 'windpipe of Eurasia'—that a British attack upon them was expected in Constantinople from the earliest weeks of the war.[36] The Ottomans thus had more than six months to strengthen their defences before Carden attacked in February, with the British bombardment of 3 November only serving to accelerate the preparations.[37] Within the Çanakkale Fortified Zone, a key figure was the commander of the Fortified Area Command (that is, the coastal defence forts and batteries), Brigadier Cevat Pasha, who was thus responsible for the successful defence of the straits. He was advised by Vice Admiral Guido von Usedom who was sent to the area in autumn 1914 with 500 German experts to hasten the strengthening of the coastal defences.[38] Only on 24 March 1915 were Liman von Sanders and the Fifth Army assigned to the defence of the straits.[39]

The Dardanelles straits are about 40 miles long and 1 mile wide at Narrows. On their northern shore, the Gallipoli peninsula presents precipitous cliffs and dominating high ground; to the south, the Asian coastline is gentler and more rolling. Fourteen permanent forts lined either side of the straits, organized into outer, intermediate, and inner defences. The outer forts at Kum Kale and Sedd-el-Bahr were very old, originating in the seventeenth century, but the inner defences were much more strongly defended, and crucially, these permanent fortifications were augmented by a howitzer zone on both shores. These howitzers, manned by the 8th Heavy Artillery Regiment, commanded by a German, Colonel Wehrle, were particularly potent weapons to fire against naval targets. The howitzers could be fired from positions that were out of reach of the flat trajectory of naval gunfire, and with their looping trajectory, their shells could fall in a perpendicular line onto a ship's decks where their armoured defences were thinner and consequently most vulnerable. By 26 February, 8th Heavy Artillery Regiment had 8,229 such shells. The plan was therefore to use howitzers to prevent ships, and particularly minesweepers, from entering the straits. For those which did run the gauntlet, there were anti-submarine nets and nine lines of mines laid across the Narrows, plus, unusually, one line laid parallel to the shore, before any enemy ship would encounter the inner defences. By 18 March, the Ottoman defences comprised 82 guns (for example, large calibre cannons) in fixed positions and 230 guns in mobile positions (i.e. howitzers).[40] But it was to be the mines which proved to be the most valuable part of this formidable set-up.

By contrast, Carden's fleet comprised a mish-mash of sixty-seven ships. There were twelve British battleships and four French battleships. Most impressively, the HMS *Queen Elizabeth* was a super-dreadnought with 15-inch guns, and hence the most powerful battleship in the world. The dreadnought battlecruiser HMS *Inflexible* was also modern with 12-inch guns, but the remainder were far less impressive. The French ships were all pre-dreadnoughts (*Gaulois, Charlemagne, Bouvet, Suffren*), and the other ten British warships were also elderly and from classes of battleship that were due to be decommissioned in a little over a year's time.[41] There were also four lightly armed cruisers (HMS *Doris*, HMS *Amethyst*, HMS *Sapphire*, HMS *Dublin*), two depot ships which played an auxiliary role, an aircraft carrier with six seaplanes, sixteen destroyers (usually used for escort duties), six submarines, one yacht, and twenty-one minesweepers.[42] The minesweepers were converted fishing trawlers manned by civilian crews. Carden's attack began in the morning of 19 February, with seven battleships firing on the outer forts. In the afternoon, the entire squadron joined in. Yet after eight hours of bombardment, no Ottoman guns had been put out of action. Due to adverse weather, the next attack only took place on 25 February.

The second day of bombardment was a lucky one, with *Queen Elizabeth* and *Irresistible* both destroying two guns early in the day. But in the following days, far more damage was wrought by small landing parties of marines who, over five days, destroyed forty-eight guns, including all the most powerful ones at the entrance to the straits. The final day of these sorties, 4 March, saw a strong Ottoman response that forced the withdrawal of the 250-strong party of marines. Meanwhile, Carden's ships had moved in to make modest attacks on the intermediate forts on 26 February, and on each of the first eight days in March. It has been calculated by historian Robin Prior that about 500 heavy shells were fired in this period, but given that only 1.6 per cent will have hit their targets, only eight heavy guns could be expected to have been destroyed.[43] It will be remembered that the Ottomans had eighty-two guns in the forts. The British simply didn't fire enough shells to have a chance of success. Even if they had fired more, the interior forts were out of their reach; and with wear and tear on the barrels, the more they fired, the less accurate they became. There was also the matter of the mines to be dealt with. After a certain amount of shadow-boxing, the unarmed, unarmoured converted trawlers, crewed initially by untrained fishermen, first saw action on 3 March, at which point they were driven away by the mobile batteries protecting the minefield. Thereafter in seventeen attempts to sweep for mines, the minefield was reached on only two occasions. Underpowered for the strong current they faced, these slow-moving craft presented easy targets in the searchlight beams for the expectant pre-ranged Ottoman guns.[44] Their advances were easily deterred.

Yet, it was not clear at first how badly the naval attack was going. Echoing the late February wave of optimism in London, there was panic in Constantinople. From January, the German ambassador, Wangenheim, had expected the British would get through. Now his Austrian counterpart, Pallavicini, made arrangements for the evacuation of his staff if there was a breakthrough. Haydarpasha railway station was thronged with people trying to make their getaway. The Ottoman government, the sultan and his court, along with their gold, cash, and most sacred relics, were also all shipped out of town to safekeeping in the interior.[45] We can gather part of an explanation for the disparity between the perception and reality of events from one of the German officers who served with the Ottoman Army. Colonel

Hans Kannengiesser later explained that the flat trajectory of naval guns sent projectiles all too often right over the top of their targets, which has a great moral effect on defenders but a limited material impact: 'This is the explanation for the almost laughably small losses on the Ottoman side, in spite of the enormous preponderance of the Entente in ships' guns which far outranged the land artillery, besides being of heavier calibre.'[46]

While the Ottomans prepared for a possible landing at Kum Kale by bringing in the 11th Infantry Division, the British decided to stage an all-out naval attack.[47] The timid Carden's health broke down at the prospect, and he was replaced by his deputy, De Robeck. On 18 March the Allied fleet, organized into three lines, entered the straits. Line 'A' was made up of the most modern ships (*Queen Elizabeth, Agamemnon, Lord Nelson,* and *Inflexible,* guarded by *Prince George* and *Triumph*), Line 'B' was made up of the French ships (*Suffren, Bouvet, Gaulois, Charlemagne*), Line 'C' was ready to replace damaged ships (*Ocean, Irresistible, Albion, Vengeance, Swiftsure, Majestic*). Line A opened fire on the intermediate and inner forts at 11.28 a.m., and was met with return fire from the forts and mobile batteries. At 12.15 p.m., the French ships of Line B moved closer to intensify the attack. An American journalist, Raymond Swing, viewed the battle from a hillside:

> The picture was in many hues, the gray-white smoke of the explosions, the orange smoke of firing cannon, and the black of flying earth in eruption, all set off by the white geysers of water as they rose after the immersion of shells. The accompaniment of sound was both oppressively insistent and varied. There was the roar when guns fired, the deafening detonations of the shells when they hit, the whistle of shells in flight, the shriek of flying splinters.[48]

A number of Allied ships were hit, including the *Queen Elizabeth,* which was struck three times, and the *Gaulois* which was forced to withdraw at 12.30 p.m. Then as Line C was ordered to relieve the French ships, one of the latter, the *Bouvet,* hit a mine and sank in less than three minutes. The captain went down with 638 of his crew. Shortly after,

Irresistible and then later *Ocean* hit mines as well. Drifting helplessly, *Irresistible* was a prime target for the forts' guns and, like *Ocean* before her, sank later that evening. Meanwhile, the French flagship, *Suffren*, was saved from a massive explosion when a young officer flooded the magazine after it received a direct hit.[49] *Inflexible* had also been badly damaged by howitzer fire and a mine. De Robeck considered the day a 'disaster'.[50] Only one heavy gun had been destroyed, no mines were swept, and a third of his squadron was out of action.[51]

By contrast, it was a triumph for the Ottomans. They still held the straits, and had suffered only ninety-seven casualties. Two episodes are particularly cherished. One is that as the French were firing on the inner defences on the European side, there was an emergency in Rumeli Mecidiye Fort after a heavy gun's loading gear was damaged after a direct hit. Corporal Seyit was inspired to lift one of its rounds on his back, so that the gun could be loaded once more. It is sometimes claimed that his shell hit the *Ocean*.[52] The story stands for the Ottoman defenders' indomitable will and strength. The other celebrated story is more important. On 8 March, the *Nusrat* had laid an extra line of mines, running parallel to the Asian shore.[53] It was these mines that sank the *Irresistible*, *Ocean*, and *Bouvet*.

The Allied plan had originally envisaged a second day of the naval assault on 19 March. With significant numbers of troops en route and their newly appointed commander General Sir Ian Hamilton just arrived in theatre, the decision was taken to postpone it until they were ready. It is sometimes argued that an opportunity to press home the attack was thereby lost and that the Ottomans were running short of ammunition, but this was not the case. More than 4,000 shells for heavy ordnance were available as well as 6,000 howitzer shells and significant quantities of other ammunition.[54] It was a clear victory for the Ottoman defenders and there was no realistic prospect of its reversal by similar means. The Ottomans had laid 402 mines up to 8 March: there was never any prospect of the British disabling them to clear a path.[55]

3

Invasion

A t Gallipoli, the armed forces of three empires squared off against each other. After the Ottoman naval victory on 18 March, a further Allied attack was fully expected. While disparate formations from the British and French empires gathered in the eastern Mediterranean, the Ottomans continued to refine their defensive arrangements. The Allies had slightly more men in theatre. They had the power of their naval guns to call upon. But they were fighting far from home, at the end of long supply lines, with half-hearted political support, and had lost almost all secrecy. Moreover, they were attacking the most strongly fortified part of the Ottoman Empire, manned by its finest forces. The invasion was launched on 25 April.

Gathering Forces

Even while the naval attack on the straits was under way, soldiers were pouring into this new theatre of war, readying themselves for battle. The men who had travelled the furthest came from Australia and New Zealand, loyal Dominions of the British Empire that had raised large volunteer armies at the outbreak of the war. Together, the Australian Imperial Force (AIF) and the New Zealand Expeditionary Force (NZEF) were formed into the Australian and New Zealand Army Corps (ANZAC). ANZAC comprised two divisions: one was the 1st Division of the AIF, an entirely Australian formation commanded by the British-born Major General Sir William Bridges, the first commandant of the Royal Military College of Australia at Duntroon; the other was the New Zealand and Australian Division, commanded by the British career officer Major General Alexander Godley. ANZAC set sail for Europe from Albany, West Australia on 1 November 1914. While they were en route, Britain declared war on Turkey, and so, rather than heading to Britain, ANZAC disembarked in Egypt to continue with their training. Lieutenant General Sir William Birdwood arrived in Egypt to assume command of the corps on 21 December. Also in Cairo was the 42nd Division, a territorial division from East Lancashire, plus the 29th Indian Infantry Brigade, which consisted of three Gurkha battalions and one Sikh battalion.[1]

The most militarily important part of the Mediterranean Expeditionary Force (MEF) was the 29th Division. Formed from garrison troops from around the British Empire, they were well-trained professionals, but had not fought together at brigade or divisional level. Two days before departure, they gained a new commander in Major General Sir Aylmer Hunter-Weston.[2] The 29th Division travelled from Britain, from its divisional headquarters in Leamington Spa to Avonmouth and then to Alexandria via Malta, and after disembarking and repacking their ships in readiness for attack, arrived at Mudros on 7 April. Equal to the 29th Division in terms of experience were the French forces of the Corps Expéditionnaire d'Orient (CEO). Its 1st Division was formed from the Metropolitan Brigade and the Colonial Brigade, and thus comprised French Europeans, Senegalese, Zouaves, and men from the Foreign Legion. It further—and unusually for Gallipoli—had the benefit of a full complement of artillery.[3] The French forces had gathered at the Tunisian port of Bizerte before beginning their transfer to Lemnos on 10 March.[4] The last part of the MEF was the Royal Naval Division (RND), a hotchpotch of naval reservists with minimal training, whose stellar ranks of gifted young men included the prime minister's son, 'Oc' Asquith, and the poet Rupert Brooke.

Opposing these forces was the Ottoman Fifth Army. Although drawn from a multi-ethnic empire, in the main its soldiers were Anatolian (that is, Turkish) Muslim peasants. But there were also

Fig. 3. The arrival of French troops in the Dardanelles.

significant numbers of Kurdish and Arab Ottomans in the army. Approximately one-quarter of graduates from the Ottoman Military Academy pre-war were Arabs,[5] and there were many officers of Libyan, Albanian, and Circassian origin. At Gallipoli, the 72nd and 77th regiments, which formed two-thirds of Mustafa Kemal's 19th Division, were predominantly Syrian Arabs. Its recruiting area also included ethnic Turks and men from the religious minorities of the Yezidi and Nusayri sects.[6] There were also Christians in the Ottoman Army (including Greek and Armenian army doctors); from April 1915 they were moved from combat to unarmed roles such as labour battalions, but this did not affect the Gallipoli campaign.

The Commanders Make Their Preparations

The commander of the Ottoman forces, General Otto Liman von Sanders, arrived at the Dardanelles on 26 March 1915 to take over command of the newly formed Fifth Army.[7] Aged 60, a man who was reserved to the point of rudeness, he had led the German military mission since December 1913. In that time, both Enver, the Ottoman minister of war, and Wangenheim, the German ambassador in Constantinople, had come to loathe him, and he had fallen out with most of the senior Ottoman officers.[8] Liman von Sanders's most important subordinate was Brigadier General Esat Pasha who had commanded III Corps since the end of 1913. His performance at the Ottoman Military Academy had been followed by three years at the Prussian Military Academy. A thoroughly professional officer and, according to Liman von Sanders, 'determined and far seeing', he was the man on the spot when the Allies attacked on 25 April.[9] On the Allied side, General Sir Ian Hamilton, commander-in-chief of the MEF, was a most charming man who enjoyed writing poetry, as well as military analysis. Of all the commanders, his career was by far the most distinguished. Now aged 62, he had had extensive experience in India and South Africa, including as Kitchener's chief of staff during the Boer War.[10] General Albert d'Amade was already in theatre, having been appointed on 2 March to the command of the CEO. Another very charming and cultured man, his nerve had already faltered in the face of the German advance on the Western Front. This new position was an unexpected second chance for which he was not well suited. He was to be recalled by early May.[11] The ANZAC commander, Lieutenant General Sir William Birdwood, had also served on Kitchener's staff in the Boer War, and was a career officer in the Indian Army. He was more successful and remained commander of the Anzacs until 1918.

Preparations to put the defence of the Çanakkale Fortified Zone (i.e. Gallipoli and the Dardanelles) on a wartime footing had begun as early as 31 July 1914. The Ottoman III Corps was allocated to this area, and it gathered in Rodosto-Tekirdağ, halfway between Constantinople and the Gallipoli peninsula by late August. Its 9th and 7th divisions were deployed to Gallipoli in mid-September and late October respectively. From 1 January, the 19th Division was assigned to III Corps and busied itself in training at Rodosto under the command of Lieutenant Colonel Mustafa Kemal Bey until early March, when it also moved to the peninsula.[12] Around the same time, as the Allied naval attacks were pressing in, the 11th Infantry Division was also deployed and held in reserve on the Asiatic shore. These divisions comprised a good number of veterans from the Balkan Wars and had been training together for as many as eight months prior to the Allied attack.[13] With the appointment of Liman von Sanders, they, together with the 5th Division, were organized into the Fifth Army. They were reinforced by the 3rd Division in early April and a new corps was activated, XV Corps, in charge of the defence of Asia, with 3rd and 5th divisions under its command.[14]

After Liman von Sanders's arrival, preparations intensified. To ensure security behind the lines, and for their own protection, the mainly Turkish and Greek population living on the peninsula were evacuated.[15] The defensive fortifications at likely landing sites were improved, as were hospital facilities; meanwhile the soldiers repeatedly practised invasion alert drills. Liman von Sanders made significant changes to the Ottoman plans for the defence of Gallipoli and the Dardanelles that had been formulated in 1912–13.[16] He studied the potential landing sites, and as a result of several useful sources of intelligence, he formed a fairly accurate idea of how many Allied troops he would face, and where and when they would land.[17] He concluded that there were two potential landing sites: the neck of the peninsula at Bulair and the Gulf of Saros, and the Asian coast at Besika Bay. He divided much of his forces between these two areas which fell under the command of German officers, but he left the peninsula itself under the command of Esat Pasha's III Corps. Esat deployed his 9th Infantry Division at the tip of the peninsula and held his 19th Infantry Division in reserve. Liman von Sanders's most important change to the original Ottoman plans (which had been based on forward defence) was the use of small platoons of men positioned near potential landing sites, ready to alert larger concentrations of men, held some miles inland who would then counter-attack.[18] When his plans were forced into action on 25 April, he was pleased with the result:

From the many pale faces among the officers reporting in the early morning it became apparent that although a hostile landing had been expected with certainty, a landing at so many places surprised many and filled them with apprehension. My first feeling was that our arrangements needed no change. That was a great satisfaction! The hostile landing expedition had selected those points which we ourselves considered the most likely landing places and had specially prepared for defense.[19]

Meanwhile, General Sir Ian Hamilton had been given the command of the MEF by Kitchener on 12 March. The next day he crossed the English Channel to France and took a train to Marseilles, where he boarded a destroyer which brought him to the island of Tenedos on 17 March. Despite Hamilton later famously claiming, 'The Dardanelles and the Bosphorous might be in the moon for all the military information I have got to go on',[20] he had a decent supply of maps made available to him in London or shortly thereafter.[21] More seriously, this hasty despatch also applied to the forces he was allocated, and precious time and secrecy were lost when they arrived in the eastern Mediterranean, for all the transport ships had to be unpacked and reorganized at Alexandria in preparation for their subsequent role in the fighting. Arriving in theatre just in time to witness the culmination of the naval attacks on 18 March, Hamilton agreed with De Robeck that the attacks should be suspended until such time as his forces could assist. This was the final step in the transition from a plan to attack by ships alone to a primarily military attack.

In assessing how best to deploy his forces, Hamilton ruled out an invasion of the Asian coast for fear of opening up a flank that could be readily attacked. The narrow neck of the peninsula from Bulair on the Gulf of Saros was also ruled out. Two diversionary attacks were scheduled for these locations: a substantial one on the Asian coast to be conducted by the French, and a more limited bluff nearer Bulair by the Royal Naval Division. Instead, mindful of the relatively small beaches on the Gallipoli peninsula, Hamilton devised a scheme focused on five beaches at the tip of the peninsula at Cape Helles. There was also an attack at dawn at a point further up the Aegean coast, labelled Z beach. This was allocated to the inexperienced Anzacs, whose task was essentially to assist the crucial landings further south, by cutting north–south communications on the peninsula and hence the flow of enemy reserves moving southwards or retreating forces moving north.[22] Overall, Hamilton's objective was the capture of the Kilid Bahr Plateau, the high ground from which the Narrows' defences could be dominated. This was never achieved.

Anzac: 25 April 1915

The Allied invasion of the Gallipoli peninsula took place on 25 April 1915. Its exact timing was a matter of some debate—should the forces take advantage of darkness to effect a surprise, or did the complexity of the operations require daylight to ensure success? Birdwood, the ANZAC commander, decided that his relatively lightly defended sector could sacrifice the support of naval gunfire for surprise and he scheduled the landings at Z beach for dawn. Hunter-Weston, commander of the 29th Division, chose to land in daylight. The risk inherent in Birdwood's choice became immediately apparent when his forces landed 1 or 1.5 miles further north than anticipated. Rather than landing on the broad Brighton Beach (north of Gaba Tepe), the forces landed in a more narrow area called Ari Burnu, which was quickly dubbed Anzac Cove. This was probably due to the difficulty in finding the correct headland in darkness. The fact that the Admiralty also had faulty maps that did not extend far enough north must also have played a part. Some have argued that the strong current also had an impact. The 'wrong' landing place has been the subject of fierce criticism in later years, but contemporaries were more sanguine. Not only was the identification of the target landing site somewhat loose and therefore flexible, but the steep cliffs surrounding the more surprising location of Anzac Cove provided some cover, while the originally intended landing site was more heavily defended.[23]

What came to be known generally as 'Anzac', the interior of the peninsula in the vicinity of Anzac Cove, is dominated by a chain of three high spots that comprise the Sari Bair Ridge: Koja Chemen Tepe (Kocaçimentepe, Hill 971), Hill Q, and Chunuk Bair (Conkbayırı). From this high ground, three ridges fan out towards the coast. This is very broken ground, covered in dense scrub, and characterized by narrow ridges and plunging cliffs leading to deep valleys. While GHQ had envisaged the target of the landings to be the capture of significant parts of the Sari Bair Ridge, before pushing inland to Mal Tepe, a high point on the southerly right flank, Birdwood considered that the three ridges should be his first objective. In doing so, he left open the risk that the Ottomans could seize the high ground to the north.

The first three waves of the landings were carried out by the 3rd Brigade, followed by the 2nd and 1st brigades of the 1st Australian Division, while the New Zealand and Australian Division had to wait until mid-morning to join the fray. In the dead of night, the first contingent of 1,500 men from 3rd Brigade left Mudros harbour on Lemnos on three battleships, and then 5 miles from the coast transferred into rowing boats that were towed twelve-apiece by steam pinnaces until they were close enough to row to shore.[24] The first wave comprised two companies each from the 9th, 10th, and 11th battalions. Lieutenant Colonel Stanley Weir, commander of the 10th Battalion, later described the experience:

> Absolute silence was maintained by all in our boats and directly the boats were cast off by the steamers we quickly rowed towards the shore. Dawn was just breaking, 4.15, and no sound was heard except the splash of the oars. We thought that our landing was to be effected quite unopposed, but when our boats were within about 30 yards of the beach, a rifle was fired from the hill in front of us above the beach, right in front of where we were heading for. Almost immediately heavy rifle and machine-gun fire was opened upon us, we had to row for another 15 yards or so before we reached water shallow enough to get out of the boats, this was about 4.15.[25]

A number of men were shot before they reached the shore. When they got there, their comrades leapt out and dashed to the beach. Many of them dumped their kit bags and pushed forwards into the scrubby brush. In these circumstances, it was inevitable that the men of the 9th, 10th, and 11th battalions got mixed up, but that was rectified to some extent when they reached Plugge's Plateau, the summit of the First Ridge. Within about fifteen minutes, 4,000 men landed at Anzac Cove. The second wave of forces, 2nd and 1st brigades, started to arrive from 6 a.m., their progress having been delayed by the need to disembark the wounded from the returning tows.[26]

The first shots opposing the landing were fired by the men of the 2nd Battalion of the 27th Regiment who were defending the coastal beaches. The gathering of enemy ships just offshore was spotted and reported by Captain Faik of 4th Company, 2nd Battalion at around 2.30 a.m. One of his comrades from the 2nd Battalion was in a trench on Hell Spit, the promontory at the southern end of Anzac Cove. He was awoken with the news:

> I was still asleep. It was before morning and the corporal who was the sentry started shouting, 'There's something unusual. Get up!' Then the company commander ordered us all to move up into the trenches. There were very few of us in the detachment, about seventy, that's all. The sentry pointed down towards the beach and we saw there were lots of them pouring out of their boats. We opened fire and they dropped down on the beach with the guns in their hands.[27]

In response to such information, Lieutenant Colonel Sefik, commanding the 27th Regiment, ordered his infantry and artillery forces 'to throw the enemy into the sea'. By 8 a.m., two battalions of Sefik's men plus artillery were en route for Anderson's Knoll (Kavak Tepe) on Gun Ridge (the Third Ridge). And well before then, from 5.55 a.m., his superior, Colonel Halil Sami, commander of the 9th Infantry Division, had set about coordinating his division's response with that of the 19th Division.[28]

For the Anzacs, it was exceptionally difficult to maintain a semblance of organization as the fighting progressed through the day. There were two key reasons for this, one topographical, the other professional. The fissured nature of the landscape meant it was difficult to advance in an orderly fashion, or to stay in touch with comrades. Small random groups of men developed. But the problem was also professional. Their limited training and inexperience, particularly among junior officers, told. These small groups emerged where battalion discipline should ideally have been maintained because of weak leadership.[29]

The threat to these forces came from the north and the south. Colonel Ewen Sinclair-MacLagan, general officer commanding (GOC) of the 3rd Infantry Brigade, chose to reinforce the right flank to guard against an attack from the south. This rendered the centre and left flanks exposed. Thus with 9th and 10th battalions already having headed south from Plugge's Plateau towards Gun Ridge and only 11th Battalion advancing northwards towards Battleship Hill (Big 700, along with Baby 700 one of two smaller hills in front of Chunuk Bair),[30] Sinclair-MacLagan, mindful of Birdwood's concern to secure the three ridges first, also ordered Brigadier M'Cay of the 2nd Brigade to further strengthen the right flank.[31] Sinclair-MacLagan also scaled back Birdwood's plans and decided against attempting to capture the Third Ridge, Gun Ridge, that day—a substantial Ottoman force had been spotted gathering there.[32] They were 2,000 men from the 9th and 19th divisions stationed at Maidos.

Meanwhile, at 5.30 a.m., Sami's counterpart, the aggressive and talented Lieutenant Colonel Mustafa Kemal, commander of the 19th Division, sent forward reconnaissance to the high ground on the northern left flank of the Anzacs' attack, that is, to Koja Chemen Tepe. With no orders having been received from III Corps by 7 a.m., Kemal decided to order his 57th Regiment along with a mountain howitzer battery to this high ground, and accompanied them there himself. Between 9 and 10 a.m., both Sefik's 27th Regiment and Kemal's 57th Regiment, heading to the south and north respectively, met up with the Ottoman forces who had been defending the coast and who were staging a fighting retreat ahead of the Australians. Kemal and Sefik coordinated their counter-attacks between 12.30 and 1 p.m., supported by three artillery batteries. This maximized the

Fig. 4. Lieutenant Colonel Mustafa Kemal (front left).

effectiveness of the outnumbered Ottoman forces (four battalions faced eight or more Australian battalions), and brought the Australian advance to a halt on both flanks. Kemal ordered a second counter-attack around 3.30 p.m., with a further attack on the Australian's right flank an hour later.[33] The fighting was ferocious and the line was pushed back and forward all day. By 10 a.m. on the northern left flank, for example, Baby 700 was captured and lost once more in a battle between the Australians and the Ottoman forces who had been defending the coast line at the outset. It was then regained, but was lost for good after Kemal's late afternoon counter-attack by the 57th Regiment, before sufficient numbers of reinforcements from General Godley's New Zealand and Australian Division could arrive.[34] It was in the course of this critical fight that Kemal gave his famous order, 'I do not expect you to attack, I order you to die! In the time which passes until we die, other troops and commanders can take our place!'[35]

It has been calculated that, with forces flowing in on both sides, ten Ottoman battalions were now facing eighteen on the other side.[36] The Ottomans' success despite this numerical inferiority had much to do with the failure by Birdwood and Sinclair-MacLagan to prioritize the high ground to the left flank. There was a good deal of dislocation and exhaustion in the ANZAC forces, as elsewhere at Gallipoli on this most difficult day. A certain number had retreated from the front line and 'straggled' back to the beach.[37] The demoralized ANZAC commanders, despite the success of landing 20,000 men, feared they would be overrun at daybreak the following day and seriously considered evacuation.[38] Both Godley and Sinclair-MacLagan expected an attack from the north. The question of evacuation passed up the chain of command, first to Bridges, then to Birdwood, who turned to Hamilton. In his only decisive contribution that day, Hamilton, having discussed the matter with his chief of the General Staff, Braithwaite, and his naval colleagues De Robeck, Thursby, and Keyes, decided against. If things were chaotic on the peninsula, how could an orderly evacuation under cover of night possibly be effected? The message went back:

> Your news is indeed serious. But there is nothing for it but to dig yourselves right in and stick it out. It would take at least two days to re-embark you as Admiral Thursby will explain to you. Meanwhile, the Australian submarine has got up through the Narrows and has torpedoed a gunboat at Chunuk. Hunter-Weston despite his heavy losses will be advancing to-morrow [sic] which should divert pressure from you. Make a personal appeal to your men and Godley's to make a supreme effort to hold their ground.
>
> (Sd.) Ian Hamilton
>
> P.S. You have got through the difficult business, now you have only to dig, dig, dig, until you are safe. Ian H.[39]

This was entirely sensible advice, but also redolent of Hamilton's prewar emphasis on morale over technical prowess.[40] The Anzacs dug in and developed a system of trenches in the coming days. The battle had become static. Meanwhile, the Ottomans were not without their own difficulties. Kemal decided to press his advantage and to continue to harry the enemy with bayonet attacks through the night. But in the process the mainly Arab soldiers of the 77th Regiment lost discipline, began to fire wildly, and were unable to press on with an attack on 400 Plateau, on the Second Ridge. They were no doubt exhausted and operating in very difficult terrain, they no longer had artillery support (it being impossible to effectively sight targets and avoid collateral damage), and many of these soldiers could not speak Ottoman Turkish—how could they then follow verbal commands in darkness? Nonetheless, some officers of III Corps formed the opinion that these 'Arab' soldiers were ill-trained and panicky.[41] Even so, the Ottomans held the high ground, and with reinforcements flowing into the area, this remained the case for the rest of the campaign.

Cape Helles: 25 April 1915

Although it is the most famous element of the Gallipoli campaign, and indeed the most important day in Australian and New Zealand military history, the fighting at Anzac Cove on 25 April was not the most important part of the invasion plan. Rather, the main effort of Hamilton's attack was focused on the toe of the peninsula either side of Cape Helles. This is indicated by the fact that his best forces, the professionals of the 29th Division, were allocated to this sector. These forces landed at five different points, labelled S, V, W, X, and Y beaches. Of these, V and W were central to the attack, and S and X were intended to guard the flanks of these attacks. Y beach, some distance northwards up the Aegean coast, was designed to threaten the flow of reinforcements or retreating Ottoman troops.

The 29th Division met with varying success on 25 April 1915. At the most easterly of the beaches, S (Morto Koyu), on Morto Bay, the South Wales Borderers found only one Ottoman platoon in opposition, and with HMS *Cornwallis* in close attendance, the battery above this wide beach was captured with only sixty-three casualties.[42] At X beach (İkiz Koyu), the westerly location designed to protect the other flank of the centre of the attack, the Royal Fusiliers, Inniskilling Fusiliers, and Border Regiment found minimal opposition and were closely supported by Captain Lockyer's HMS *Implacable*. Strong attacks later in the day very nearly succeeded in driving the invaders into the sea, but again naval fire support helped to see off the threat. At the more distant Y beach (Zengindere), 25 and 26 April were a rather curious couple of days. Without specific orders or strong leadership from Lieutenant Colonel Matthews of the Royal Marines, the King's Own Scottish Borderers lacked drive and direction. They eventually entrenched not far from the beach, and after holding off a night attack from the Ottomans, they re-embarked on the 26th citing lack of reinforcements and ammunition, a process that may well have begun without Matthews's knowledge or orders. Thus a mixture of success, minimal casualties, and, in the case of Y beach, near-farce, characterized the three peripheral beaches. At V and W beaches, however, it was a quite different story. Here the British attacked difficult and strongly defended beaches which were manned by well-trained and tenacious Ottoman troops. The result was ferocious fighting during which the British struggled to land on or move beyond the beaches.

The defence of the Cape Helles area was the responsibility of the Ottoman 26th Regiment which had been stationed there since 19 August 1914. It had thus spent eight months in improving the defences and honing and coordinating its plans. When the 25th Regiment also arrived in the area as reinforcements, Colonel Sami was able to allocate the entire 26th Regiment to the defence of these beaches on 21 April 1915. The 1st Battalion was allocated to the Kum Tepe defences on the Aegean coast (i.e. just north of Y beach). The 2nd Battalion was held in reserve. This left Major Sabri and the thousand men of the 3rd Battalion in charge of the defences at Sedd-el-Bahr. Sabri allocated two platoons from 10th and 12th companies each to X, V, and W beach, and held his other two companies, the 9th and 11th, in reserve. In addition, the Ottomans had forty-four artillery pieces in this southern sector, plus three howitzer batteries from the Strait Fortress Command.[43]

The Ottomans became aware that an attack was imminent from 23 April, and the platoons at the beaches were ordered to open fire when the invasion forces came within 200–300 yards of land. The Ottomans' growing awareness of Hamilton's plans is reflected in a stream of reports from the beaches back to regimental and corps HQs. At 3.20 a.m. on 25 April enemy activity was spotted off Tekke Burnu (i.e. the northern end of W beach). At 4.30 a.m. the Royal Navy began its bombardment, and at 6 a.m. dozens of boats were seen approaching both V and W beaches. By 7 a.m., it was clear that significant landing forces were targeting these locations. In particular, the approach of the steamer *River Clyde*, with hundreds of soldiers on board, was interpreted as a sign of the importance of V beach as a landing site and triggered Sabri to release half of his battalion reserve, 11th Company. (The other half of his battalion reserve, 9th Company, was despatched to X Beach before 10 a.m.)[44]

Fig. 5. V beach, Cape Helles, photographed during the Australian Historical Mission's return to the peninsula in February and March 1919. The ship to the left was HMT *River Clyde* which was run aground on 25 April 1915 to enable Irish and British soldiers to get closer to the shore.

The use of the *River Clyde* in the landing, as a modern-day Trojan Horse, was a response to the defensive strengths of V beach (and to a shortage of small boats). Often described as a natural amphitheatre, its broad, sandy beach was ringed by high ground linking two headlands. The easterly side was dominated by the Sedd-el-Bahr fortress. To the west lay Fort Ertrugal (Fort No. 1, the Ottomans called V beach Ertuğrul Koyu) and, behind it, two small hills, Guezji Baba (Gözcübaba Tepe) and Ay Tepe, and behind them, Hill 138. The plan was that, preceded by a naval bombardment, most of the attacking force of the Royal Dublin Fusiliers were to approach the beach in wooden boats towed by steam pinnaces at 5.30 a.m. Thirty minutes later, the Royal Munster Fusiliers, more Dublins, and some of the 2nd Hampshire Regiment were to emerge from the *River Clyde* using a bridge of small boats to reach the shore. The reality was carnage. The timings went awry and the naval gunfire had only a limited effect. The Ottoman riflemen stationed in the trenches and other defences surrounding V beach trained the fire on the incoming boats with devastating effect.[45] Many were killed in the boats; others who were wounded, drowned. Those waiting to jump out of the sally ports cut in the side of the *River Clyde* found that the bridge of boats had not been put in place. To rectify matters, Commander Unwin and Able Seaman Williams jumped into the water themselves to pull two boats into place so that the troops could reach dry land. Even with this improvisation, reaching the shore was a deadly task. Captain Geddes of the Munsters tried to lead his company to the beach. With machine guns trained on the exit points of the ship, he later reported:

> We got it like anything, man after man behind me was shot down but they never wavered. Lieutenant Watts who was wounded in five places and lying on the gangway cheered the men on with cries of 'Follow the Captain!' Captain French of the Dublins told me afterwards that he counted the first 48 men to follow me, and they all fell. I think no finer episode could be found of the men's bravery and discipline than this— of leaving the safety of the *River Clyde* to go to what was practically certain death. I dashed down the gangway and already found the

lighters holding the dead and wounded from the leading platoons of 'Z' Company.[46]

The small numbers of men who did reach the shore found only a low sandy bank to protect them.[47]

The situation at V beach remained desperate until at least mid-afternoon, when British forces from neighbouring S and W beaches began to enfilade the Ottoman defensive positions there. With Colonel Sami, in one of his few errors, hesitating to commit his regimental reserves during the afternoon, one of the Ottoman strongpoints above V beach, Ay Tepe, was lost at 5.40 p.m.[48] Then, during the evening, their remaining troops having finally been able to disembark around 8.30 p.m.,[49] the Irish troops turned their attentions to the other strong-point above V, Guezji Baba. It was fiercely defended by Sergeant Yahya of 12th Company, 3rd Battalion, 26th Regiment who stepped into the breach to lead sixty-three men. When the Irish finally broke into his defences, Yahya led a bayonet charge to drive them out once more, before in turn being driven out under heavy fire. Then, in the small hours of 26 April, the Ottoman regimental reserve joined the fray. The 1st Battalion of 25th Regiment, in combination with the survivors of 9th and 12th companies of the 3rd Battalion, renewed the attack on Ay Tepe at 3.30 a.m. At the same time, 11th Company attacked Guezji Baba. But by this time, the Irish were able to bring superior firepower to bear and the attacks were beaten off after an hour.[50] They had secured their landing place, but had not progressed beyond it. The multi-brigade spearhead of the Allied attack had been held up by one Ottoman infantry company.[51]

The story of W beach on 25 April is scarcely less bloody. W beach (Tekke Koyu) is narrow—only 200 yards wide—and overlooked by cliffs. It was defended by trenches and wire, and 200 Ottoman men from the 12th Company of 3rd Battalion, 26th Regiment. It fell to the 1st Battalion, Lancashire Fusiliers to attack here, and subsequently the beach became known to the British as Lancashire Landing. As elsewhere, the preceding naval bombardment left the defenders relatively unscathed, and the Ottomans opened fire on the approaching thirty-odd rowing boats, towed in lines by steam pinnaces, when they neared the shore. Again, many were killed or wounded on board, and it was initially impossible to make any progress from the water's edge. Only when the focus of the second wave of the attack was switched to the left-hand side of the beach, where the climbable cliffs provided some cover, were any men able to disembark. Led by Major T. Frankland, about 100 men proceeded back along the clifftop towards V beach, assisted by the fact that one Ottoman unit had been diverted to defend X beach. Although Frankland was shot during his advance, his initiative, reinforced by the 4th Worcester Regiment and the 1st Essex Regiment later in the day, secured W beach and the link made with the adjacent V beach.

The British captured the fourth and final hill surrounding Cape Helles on the morning of 26 April. A concerted attack saw the Hampshires fight their way through Sedd-el-Bahr fort and village, and thereafter Hill 141 was captured following a charge led by Lieutenant-Colonel Doughty Wylie, who died stick in hand at the head of his column.[52] But, beyond the captured area lay Achi Baba, the hill that dominated the toe of the peninsula. It was supposed to have been won by the end of the day, prior to a push on the Kilid Bahr Plateau on 26 April, but remained untouched.

Diversions

There were also two diversionary attacks targeted at exactly the remaining spots where Liman von Sanders had anticipated landings. At the Gulf of Saros, the RND put on a show of preparing to land during 25 April, and then that night, in a daring act of individual heroism, the New Zealander Lieutenant Commander Bernard Freyberg swam ashore to light a series of flares in a bid to suggest imminent invasion. Remarkably, he then managed to swim back out to sea and locate one of the awaiting rowing boats.[53] But his action does not seem to have deceived the Ottomans for long, and Liman von Sanders ordered the 7th Infantry Division, which had been positioned to defend Bulair, to march southwards to provide reinforcements during the late afternoon of 25 April. Nevertheless, partly as a result of intelligence gained from Lieutenant Palmer who had been captured from a submarine, E 15, which sank in the straits on 17 April, Liman von Sanders remained nervous about the threat at the Gulf of Saros and kept the 5th Division there until 28 April.[54]

The second diversionary attack was a far more significant undertaking. At the edge of the Asian mainland lies a strip of low-lying land separated off by the Mendere River which runs parallel to the Aegean Sea. At the Dardanelles end of this strip is the Kum Kale fort and at the other, the village of Yeni Shehr (Yenişehir), whose gun installation had already been destroyed back in February. These positions were won relatively quickly by the French 6th Colonial Regiment which had landed almost unopposed on 25 April. The narrow strip had been judged to be impossible to defend due to its vulnerability to naval gunfire. But across the river lay the Ottoman 3rd and 11th divisions, ready to counter-attack that night. The German commander of the 3rd Division, Colonel August Nicolai, encountered significant difficulties in tactically coordinating his forces, in contrast to the more well-rehearsed forces of the 9th and 19th divisions that faced the Anzacs. Nonetheless, a bloody night of attack and counter-attack ensued, with terrible losses inflicted on the predominantly Senegalese attacking force and the defending Ottomans. On 26 April Hamilton called for d'Amade's forces to be transferred to Helles, and so the diversionary attack came to an end with a complete withdrawal by the end of the day. D'Amade's men had suffered 768 casualties, while the Ottomans had lost 1,735 men, with a further 500 taken prisoner.[55]

A Bloody Reckoning

The multiple landings on and near the Gallipoli peninsula on 25 April were a remarkable feat of heroism and organization. More than 200 ships were involved in getting the expeditionary force into position.[56]

Perhaps 20,000 men landed at Anzac Cove[57] and 17,000 at Cape
Helles that first day.[58] They significantly locally outnumbered the
defenders on the peninsula. The ratio of Allied to Ottoman forces at
the landing was something like 1.6:1, and by the end of the first week,
when the French forces had been transferred over from Kum Kale,
the gap widened to 2:1. Expressed in terms of a small army unit, it
has been calculated that forty-four British companies attacked twenty-
four companies at Cape Helles, and that forty-eight ANZAC companies
attacked twelve companies at Anzac Cove, until the inflow of Kemal's
reserve 19th Division raised the Ottoman total to forty-two.[59] The
Ottomans lost 10 officers and 1,887 men in the Cape Helles sector facing
the British landings. British casualties numbered 3,800 on 25 April.[60]
Equivalent statistics for the first day at Anzac Cove are harder to come
by. For the nine days to 3 May 1915, casualties at Anzac have been
estimated at 8,500 Australians and New Zealanders, 600 British, and
14,000 Ottomans. At Helles, the British lost 4,453 casualties by 30 April,

Fig. 6. Anzac Cove, 25 April 1915. The beach became very crowded following
the landing, not least because insufficient arrangements were made for coping
with the dead and wounded.

the French 1,779.[61] By 30 April, the 1st Australian Division (not the entire ANZAC) had casualties of 4,931.[62]

The MEF had achieved a feat unprecedented in modern warfare with their amphibious landing in the face of determined opposition. They did not, however, achieve their goals for the first day, and wobbled on the brink of a disastrous withdrawal in one sector. The historian Ed Erickson has drawn a useful distinction between the two main areas of activity which helps us to understand what went on:

> Whereas the fighting at Anzac on 25 April proved to be essentially a movement to contact resulting in a meeting engagement followed by hasty Ottoman attacks, the fighting at Cape Helles was characterised by direct British assaults on an enemy strongpoint system. It was more like the fighting then raging in France and, consequently, was far more violent, resulted in far more British casualties and also culminated in a large number of Victoria Crosses being awarded for acts of gallantry.[63]

Further differences apply if we consider the artillery support available to the infantry at the landings. In historian Tim Travers's view, naval cooperation was the decisive factor at the Helles landings. In the face of strongly entrenched defenders, naval firepower was crucial and where individual ship's commanders decided to approach as close as possible, its impact told (as with *Implacable* at X beach). Concomittantly, its absence or ineffectiveness cost many lives—such was the opinion of an irate Commodore Keyes. At V beach, *Albion* stayed too far out and *Queen Elizabeth*'s awesome firepower was not properly brought to bear.[64] At Anzac, the Ottoman artillery had a murderous effect, while the Anzacs found it almost impossible to find a suitable position for their guns in the terrain. The 26th Indian Mounted Battery did manage to haul six guns into position on top of 400 Plateau during the afternoon of 25 April, but they were soon forced back.[65]

A comparison of the professional readiness of the soldiers on opposing sides is also instructive. The Ottoman Army conducted a skilful and determined defence of their homeland, due in large measure to its officers. Vastly improved in quality since the Balkan Wars through training, vigorous debate, and the retirement and replacement of older officers, they had a decisively positive impact.[66] III Corps was the best formation of the Ottoman Army, and unchanged in its organization since fighting in the Balkan Wars, experienced, and well drilled. The Ottomans outfought their opponents, maintaining a determined defence from superior positions on higher ground combined with a series of well-coordinated counter-attacks. The average Ottoman soldier—known as Mehmetçik—proved to be a tenacious defender of his homeland. What the Ottoman Army did lack, however, were experienced non-commissioned officers. Vital to the dynamics of smaller units, their absence helps explain why the Ottomans did better in defence than in attack.[67] By contrast the ANZAC corps had had limited training and no prior experience of warfare. In such circumstances, individual bravery and initiative are of minimal compensation. Similar issues of inexperience, including its leadership, applied to the RND.[68] Furthermore, the MEF demonstrated significant deficiencies in command and control. Hamilton has been criticized for his stance in leaving his plans to unfold without intervening, save for when his opinion was asked regarding evacuating Anzac. In particular, he did not press Hunter-Weston to intervene at Y beach, and Hunter-Weston erred more generally in attacking at the strongest point (V and W) rather than exploiting weakness.[69] By contrast, Liman von Sanders and Esat were decisive and flexible in the ways in which they committed reserves, and reorganized command according to developments on the ground.[70] In terms of leadership on shore, it can be seen that where it was largely absent, as at Y beach, chaos ensued. Where it was misguided, as in Birdwood's and hence Sinclair-MacLagan's failure to prioritize high ground, tactical advantage was ceded to the defenders. Where it was brilliantly bold, as in Mustafa Kemal's intervention, it was decisive. Through his actions, the Ottomans saved the high ground, and thus held the advantage throughout the ensuing campaign. The threat from the MEF had been held at bay by a tenacious, flexible, and well-organized Ottoman defence.

4

Stalemate

By the first week of May, the two beachheads at Gallipoli were established but the decisive blow had not been struck. The Ottomans had not driven the invaders back into the sea during the landings. The Allies had not captured the high ground on the peninsula, which would enable the opening-up of the Dardanelles straits. Instead, after their extraordinary exertions and heroics, the exhausted forces regrouped. At Helles they reorganized to incorporate the French into the right of the line, while at Anzac, they dug in to protect the small area they had won. In the following weeks, the main focus of Mediterranean Expeditionary Force (MEF) activity was at Helles.

Krithia

The Allies launched three major offensives at Cape Helles between late April and early June in an attempt to capture Achi Baba. Each failed horribly due to hurried planning, poor artillery support, and unimaginative and overambitious tactics. The attacks and the Ottoman response were as follows:

- First Krithia (28 April). The exhausted men of three Allied brigades led by inexperienced officers were supposed to capture positions that would improve the chances of a subsequent attack on Achi Baba. The 87th Brigade on the left was expected to advance 5 miles. The 88th Brigade, which started the battle in some disarray, was to attack in the centre, and, with the French

175th Brigade on the right, they were to wheel around to the right to capture the village of Krithia and a line almost perpendicular to their starting point. With artillery support that was only 10 per cent of the quantity that would have been used on the Western Front at this time, and no means for these heavy guns to move forward with any advance, this battle across open country was over before the end of the afternoon.[1] Although the Ottoman 9th Division on their eastern flank came close to withdrawing around noon, the stalwart regimental commanders held their nerve and instead counter-attacked, thereby thwarting the Allied advance.[2] Allied losses were about 3,000, and no ground was gained.[3]

- Ottoman night attack (1/2 May). In the early hours of 1 May, Kemal had led an attack that the Royal Naval Division (RND) just about managed to hold off. Then, that night, using newly arrived reinforcements and all three divisions at Cape Helles, further attacks were launched in the southern sector. These attacks failed because there were temporarily too few staff officers in the HQ of the newly formed Southern Group. They therefore couldn't react sufficiently swiftly to the reports of the battle as they came in, and allocate reinforcements appropriately. A further attack on 3 May also suffered from severe command and communications problems at HQ. More than 10,000 casualties were suffered over four days.[4] With the loss of over 60 per cent of the officers in the 15th Division, and similarly in the 2nd Division in the 19 May attacks, the usefulness of these Ottoman formations was destroyed.[5]

- Second Krithia (6–8 May). This was another overly complex Allied attack right across the front line, aiming to capture Achi Baba in three stages. It involved, from left to right: fresh arrivals from 42nd Division (125th Brigade), the 87th and the 88th Brigade—a composite division which included an Australian and a New Zealand brigade—the 29th Indian Brigade, and the French Corps Expéditionnaire d'Orient (CEO) with some RND reinforcements on the right. With inadequate artillery preparation firing on indeterminate Ottoman targets, three terrible days followed. Complex but unduly brief orders were hurriedly issued to very tired men. Rather than calling off the battle, Hamilton decided on the final day to call for movement en masse across the line with bayonets fixed. These were ridiculously unsophisticated tactics unsuited for modern warfare.[6] Throughout the battle, the Ottomans held several regiments in reserve and their overall position was never seriously under threat. The Allies suffered 6,500 casualties; the Ottomans, 2,000.[7]

– Ottoman night attack (18/19 May) at Anzac. A disastrous attack that signalled the end of large-scale offensives by the Ottomans in this sector.[8]

– Third Krithia (4–6 June). The last large-scale attack at Cape Helles in this phase was another division-scale operation. On the left of the line the Indian Brigade was to attack at Gully Ravine (Zengindere), to their right were the 88th Brigade, the 127th Brigade, and 125th Brigade, then the Royal Naval Division, and the French who were detailed once more to attack Kereves Dere Ravine. Hunter-Weston's plans were more realistic this time, and the artillery barrage was more sophisticated. Yet the fusillade still lacked sufficient weight and precision, and, crucially, stopped two minutes before the infantry attack was to begin.[9] On the right, the French fared particularly badly, as did, in turn, the Collingwood battalion of the RND when the French attack stalled. There was briefly some progress in the centre with, at one point, a 1,000-yard advance by the 88th Brigade,[10] and some temporary success for the Sikhs at Gully Ravine. Yet after fierce fighting the Sikhs were forced back to their start line. The Allies suffered 4,500 British casualties and 2,000 French,[11] for an advance of 500 yards in the centre of the line. Although the 12th Division reported that morale was 'shaken', at no point did the Ottomans have to commit their divisional reserves to the fray. There were 5,017 Ottoman casualties, of whom 52 were killed.[12]

Fig. 7. Resting soldiers at Gallipoli. There was very limited space to relax behind the lines at Gallipoli. These men are sheltering under the cliffs between Cape Helles and Gully Ravine as a shell from Asiatic Annie bursts in the sea.

After another Ottoman counter-attack on 6 June which brought further heavy losses to all concerned, there were no further large-scale attacks at Cape Helles until late June. From that time on, Hamilton and Hunter Weston opted for smaller divisional-scale attacks that adopted the less dramatic but more realistic tactics of 'bite and hold', which had more modest goals and much more effective artillery bombardments preceding them.[13] These were the French attack on Kereves Dere of 21–2 June and the 29th Division's attack on Gully Ravine (28 June–5 July). The Ottomans counter-attacked at Gully Ravine on 5 July with their most sophisticated use of artillery to date, yet the result was the loss of some 16,000 Ottoman troops.[14] A second Allied attack on Kereves Dere (12–13 July) also demonstrated some benefits from strong artillery preparation, but at heavy cost to strained and often sickly men. Thereafter, with the departures of the authors of these 'bite and hold' attacks—Hunter Weston (sick) and General Henri Gouraud (wounded)—from the peninsula, plus the developing option of a renewed manoeuvre battle, limited attacks and steady progress were abandoned for the remainder of the campaign.

Sanitation and Hospital Provision

After Second Kereves Dere, morale plummeted. This was partly due to huge frustration with Hunter Weston's muddled orders (which seriously undermined the fact that he essentially had the right idea). But there were also two other significant problems that had emerged. The first was the arrival of German U-boats in the area, which caused the withdrawal of the larger support ships. HMS *Goliath* had already been sunk by an Ottoman torpedo boat at the mouth of the straits. As it went down, one of its officers ordered the crew, 'Keep calm, men! Be British!' Then on 25 May, the *Triumph* was sunk off Anzac, and on the 27th, the *Majestic* met the same fate off W beach. These losses added to a sense of vulnerability for the men clinging on to the peninsula.[15] The second horrible problem to be faced was the state of the battlefield, which was becoming congested with dead bodies. This was an issue across the peninsula, and had significant effects on hygiene arrangements and sickness rates in consequence. It became most acute in mid-May at Anzac after the extensive losses in the Ottoman attack of 19 May, and necessitated an armistice to clear the battlefield.

The task of getting casualties away from the battlefield was particularly difficult at Gallipoli, hence the ambulance man Simpson's ad hoc arrangement with a donkey. Even once the wounded had reached the beaches, they were vulnerable to enemy fire, and a shortage of transports made it difficult to evacuate them in a timely fashion, particularly in the first days after the landings. The numbers of expected casualties had been grotesquely underestimated. Even the anticipated 10,000 casualties, 2,000 of whom were expected to be serious, would have overwhelmed the three hospital ships. The seven hastily

Fig. 8. A dramatic image of HMS *Majestic* sinking on 27 May 1915 after it was hit by a torpedo off Gallipoli.

converted transport ships were not properly fitted out for their allocated work: they weren't even identified as hospital ships to spare them from enemy fire. Such was the outflow of casualties that within the first ten days, every hospital bed in Alexandria had been filled. The root of the difficulties lay in the challenges for Hamilton's GHQ to liaise with his administrative branches—the Adjutant and Quartermaster General staff, including those whose job it was to plan the medical evacuation, were left behind in Egypt due to lack of space on Lemnos or on the ships while the landings were first being drawn up.[16]

The planning failures for the August landings were not quite so egregious, with a more realistic expectation of 30,000 casualties and a

Fig. 9. Mateship in action. An Australian soldier carrying a wounded man to receive medical attention.

better supply of medical personnel and hospital ships. More tented facilities were set up on Imbros and Lemnos, including two Canadian Stationary Hospitals from mid-August. Yet water shortages there meant that staff were unable to adequately clean the patients, the laundry, or the hospital. As on the peninsula, only a single pint of water per person was available for drinking and hygiene. In consequence, diarrhoea was rife, even among the nurses: two Canadian nurses died from dysentery and twelve out of the remaining twenty-five were invalided home.[17] Moreover, some of the evacuated casualties were found to be malnourished after the August landings, such were the limitations of their food supplies. Altogether, 1,000 men were evacuated due to disease during the campaign.[18]

By contrast, although their conditions worsened during the campaign, the Ottoman Army benefited from a field hospital for every army corps, and ready access to 20,000 hospital beds in southern

Fig. 10. Ottoman soldiers camped at Gallipoli in 1915.

Thrace and Constantinople.[19] The army was also well fed and watered—Ottoman soldiers were allocated 3,149 calories per day and enjoyed a reasonably varied diet which included 900 grams of bread each day and fresh fruit and vegetables when they were in season.[20] Nor were the Ottomans unduly troubled by sickness, the ratio of wounded to sick was 24:1 between 25 April and 1 July. The medical services seem to have been well regarded.[21]

The August Battles: Anzac

Despite these circumstances, discussions among the MEF commanders continued through May, June, and July as to how best to renew the attack. The idea of a breakout from Anzac emerged. After a period of political paralysis in London was resolved, the decision to send out two New Army divisions as reinforcements meant that plans were put on hold until their arrival. Thus in early August an audacious attempt to capture the high ground above Anzac, coupled with a new landing at Suvla Bay to secure a winter harbour, was put into action. The Ottomans were waiting for them. Rumours had been rife that a renewed assault was under preparation, but opinion differed within the Ottoman high command as to where the blow would fall. In consequence, Liman von Sanders allocated three divisions to meet an attack on the Asiatic coast at Kum Kale, and three divisions at Saros Bay in case the neck of the peninsula should become a focus. His allocations to Anzac and Cape Helles meant that the two sides would initially be roughly evenly matched, but Liman von Sanders also had six divisions en route to the peninsula and those six divisions at Saros and Kum Kale to draw upon. The Ottomans were, at first, perhaps most outnumbered at Suvla where Major Willmer commanded a group of about 4,000 men.[22] Nonetheless, Suvla, while securing some new ground, was a dismal failure. The Anzac breakout is usually regarded as a close-run thing: two out of the three high points in the Sari Bair Ridge were temporarily captured.

The renewed assault proceeded as follows. To distract the Ottoman forces from the night-time breakout attempt from the northern sector of Anzac, the attacks were preceded by diversionary attacks at Helles and at Lone Pine in the south of Anzac on 6 August. The territorial soldiers of the Manchester Regiment were in the thick of the fighting at Helles that caused 3,335 British casualties, and 7,500 Ottoman casualties by 13 August, but did not prevent the Ottomans moving forces northwards to Anzac.[23] The ferocity of the fighting at Lone Pine has gained it lasting notoriety. The 1st Australian Brigade crossed the 100 yards of no man's land at 5.30 p.m. to attack Lone Pine (Kanlisirt— which translates as Bloody Ridge). It was held by Major Tevfik's 47th Regiment. The Ottomans had covered their trenches with logs as protection from bombardment, and the Australians had to drag these off and then jump down to engage in hand-to-hand fighting and bombing within the confines of the trench system. The Australians captured the first line of trenches within thirty minutes, and then faced a counter-attack from 11 p.m. featuring Ottoman regimental and divisional reserves, which saw the death of both Tevfik and Lieutenant Colonel Ibrahim Şükrü, the commander of the 15th Regiment, which was part of the first wave of reinforcements. Further Ottoman reinforcements were drawn into the area, and as such, the diversionary attack on Lone Pine can be judged a success. Vicious fighting and counter-attacks in which the Australians won seven VCs continued until 10 August. The Ottomans lost 1,520 killed, 4,750 wounded, while the Australians lost 2,277 casualties.[24]

Meanwhile, overnight on 6/7 August two columns set out to climb up the gullies to the north of the sector to get into position to attack the isolated Ottoman posts guarding the strategic high ground. The Left Assaulting Column, consisting of the 4th Australian Brigade (commanded by General John Monash) and the 29th Indian Brigade (Major-General Herbert Cox), made their way up the more northerly Aghyl Dere in order to attack Hill 971 (Koja Chemen Tepe) and Hill Q (Besim Tepe) respectively. A covering force made up of half of the 4th Brigade preceded them. The Right Assaulting Column comprised the New Zealand Infantry Brigade commanded by General Francis Johnston; its covering forces were also New Zealanders: the New Zealand Mounted Rifle Brigade and the Otago Mounted Rifles. This column split up to make their way up Chailak Dere and Sazli Dere, before reforming in order to attack Chunuk Bair (Conkbayiri). From there, they were to move south-west towards Battleship Hill and onwards to meet up with an attack coming in the other direction across the Nek towards Baby 700, led by the dismounted 8th and 10th Light Horse regiments.[25]

These night marches through difficult ravines and nullahs (dried-up streams) without suitable maps were supposed to culminate in coordinated attacks. This would have been practically impossible even if many of the men involved were not debilitated by dysentery. Skirmishes along the route with Ottoman soldiers further added to the inevitable confusion. Impeding the Anzac Left Assaulting Column were men of the Ottoman 14th Regiment, defending the slopes of Hill 971 and Abdul Rahman Spur. Its 1st Battalion was swiftly reinforced by

Lieutenant Colonel Ali Rifat who ordered forward the 3rd Battalion and elements of the 2nd. At dawn on 7 August, most parts of the Left Assaulting Column were 3 miles from their objectives. Only 1st Battalion, 6th Gurkha Rifles of the Indian Brigade were close to their objective, Hill Q. The Right Assaulting Column was similarly dislocated, but a small group under Johnston's command were near the summit of Chunuk Bair. He decided to wait for reinforcements to join him before moving on the summit. While he did this, it so happened that Colonel Hans Kannengiesser had been reconnoitring Hill Q and Chunuk Bair. Kannengiesser had arrived at Hill Q at 6.30 a.m. with reinforcements from 64th Regiment. The 25th Regiment was not far behind; Kannegiesser ordered it and the 14th Regiment already in situ to Chunuk Bair. They were ready and waiting when Johnston's men finally attacked around 10.30 a.m. on 7 August. The Aucklands, the Gurkhas, and then the Canterburys were mown down by Ottoman fire.[26]

The attack on the Nek was also scheduled for the morning of 7 August. Initially conceived as part of a pincer movement following on from the capture of Chunuk Bair, with news of the slow progress of the Right Assaulting Column, it became a feint attack to assist the battle at Chunuk Bair, along with other feint attacks at the Chessboard, Dead Man's Ridge, and German Officers' Trench. The attack at the Nek has become notorious for the bloodshed which ensued. Four lines of Light Horsemen attacked. After the first attack was decimated by machine gun fire, it was clear that subsequent attacks would be deadly. Yet a second wave of attacks went ahead, and even after appeals to Brigade HQ to call it off, a third and a fourth. As men from the 10th Light Horse took their places for their part of the attack, their commanding officer told them, 'Boys, you have ten minutes to live, and I am going to lead you.'[27] Of the 1,250 Light Horsemen who attacked, 650 were killed or wounded.[28]

No further meaningful attacks took place on 7 August and orders were issued for a concerted attack at dawn on the 8th. These forces were exhausted, desperately scattered and mixed up, and grappling with very confusing and challenging ground. Monash's and Cox's men made no headway, but a small group of New Zealanders under Colonel Malone, along with some British troops, did capture a toehold on Chunuk Bair. However, unable to dig in at the rocky summit, enfilade fire from Koja Chemen Tepe and Hill Q and fierce counter-attacks pushed them off the summit after Malone and many of his men were killed late in the afternoon. Yet they clung on just below the summit for the rest of the night. Ottoman officers had also suffered important losses: two divisional commanders wounded and two regimental commanders had been killed.[29] One of the former was Kannengiesser who had been seriously wounded on 7 August. Lieutenant Colonel Cemil of 4th Infantry Division replaced him in command early that afternoon. Cemil was placed in charge of an ad hoc group, 'the Aghyl Dere Detachment', comprising elements from six different units. Remarkably, Liman von Sanders continued to reorganize his structures effectively as battle raged, and thus Cemil was subordinated to Colonel Ahmet Fevzi from 9 a.m. on 8 August.[30] Fevzi himself was replaced as commander of the Anafarta Group overnight on 8/9 August by Mustafa Kemal, who was far more familiar with the area.

Major General Godley, commanding the New Zealand and Australia Division, ordered renewed attacks by his exhausted men on 9 August to secure the tenuous grip on Chunuk Bair. These were to no avail. The New Zealanders themselves were relieved by an inexperienced New Army battalion, the Lancashires, around 10 p.m. that night. Then in the early hours of 10 August, Kemal's forces struck the decisive blow. With fresh troops at his command, he personally crept forward to launch sixteen battalions onward and swept the British and New Zealanders from their toeholds and removed the threat to the Sari Bair Ridge. Heavy losses were inflicted on both sides. A shrapnel ball hit Kemal himself, but his pocket watch deflected it.[31] New Zealander Private Leonard Hart of the Otago Battalion later wrote home about these events:

In case you may have heard different I may say that we never at any time held the whole of the great ridge known as Sari Bair, but the most important part of it, known as Chunuk Bair, was in our hands and it was from our position here, that, had we been able to hold it, we could have dominated practically the whole southern portion of the peninsula. Much blame and ill feeling has been created between the Colonials and Tommies over them not putting up a better fight when the Turks attacked, but I am inclined to think that, judging by the frightful losses sustained by the Wellington and Auckland battalions while holding the position, we would not have done much better.[32]

In the same action, Major Allanson and his 1st Battalion, 6th Gurkhas pushed forward at the adjacent Hill Q, but were hit by artillery fire—possibly 'friendly fire'—and had to retreat from an important tactical position. Chunuk Bair is known in Turkey as Kemelyeri (Kemal's place).[33]

The August Battles: Suvla

With battle raging at Lone Pine, and the assaulting columns making their way towards the high ground, a further element of the August offensive got under way on the evening of 6 August. This was the IX Corps' landings at Suvla Bay to the north of Anzac. The corps was commanded by Lieutenant General Sir Frederick Stopford, a man appointed purely for his technical seniority rather than his battle experience. He was old and incompetent. The IX Corps comprised the first New Army units to fight in the war—these were men who had responded to Kitchener's call at the outbreak of the war. They had been trained for the trench conditions of the Western Front, not the mobile operations now required of them. Moreover, they had little time to recover from the dysentery that had struck many of them in camp. It did not augur well.

The purpose of Suvla was to secure a harbour, and as such it was a subsidiary operation to the Anzac offensive. This helps to explain some of the reasons why the IX Corps did not push forward as vigorously as they might have done. The entire operation has been severely criticized for letting down the Anzac breakout, but in fact it was largely irrelevant to the success or failure of the early August operations there. Stopford has been scapegoated for failures for which he had no responsibility. Nonetheless, plenty of things went wrong at Suvla. Firstly, secrecy on the part of Hamilton and his staff meant that there was insufficient time for Stopford and his staff to draw up appropriate plans; the result was some hopeless muddle even before his forces departed from Imbros. Secondly, the plans were unduly cautious, shaped by the experience on the Western Front of Stopford's chief of staff, General Hamilton Reed. But Suvla was not heavily defended, and the Ottomans were not well dug in there. The plan for the landing force should have emphasized pushing onwards to capture the high ground of Kiretch Tepe and Tekke Tepe; but it did not. Thirdly, as at the April landings, there were significant communication problems that prevented the commander from following proceedings and issuing further orders. Stopford's HQ was on board the *Jonquil*. He did not have a phone line until 8 August, and relied instead on semaphore messages conveyed through one benighted soldier on the ship. Stopford went ashore a day and a half later than planned, on the evening of 8 August, and even then technical difficulties in communicating with his HQ persisted. A fourth problem was the unsuitability of Suvla Bay as a landing point. It had too many shallow sandbanks, which served to dislocate the landings, and many units had to land in unplanned locations. This caused enormous confusion in the dark.[34]

The 11th Division was entrusted with the initial landings. The 33rd Brigade and 32nd Brigade had uneventful, unopposed landings to the south of Suvla Bay at C and B beaches; they then captured nearby Nibrunesi Point and the high ground of Lala Baba in short order, but in the process all but three of their battalion officers were killed.[35] Without their leaders, these inexperienced troops decided to await further developments. Things really started to go wrong with the landing of 34th Brigade inside Suvla Bay. It proved to be a shallow and unsuitable bay for landing, with plenty of low sandbanks just below the water. Various landing craft became stuck, and the troops on board made their way ashore in commandeered small boats. As a result, they became horribly mixed up, as did two battalions which managed to land at the beach, and further confusion ensued as the battalion which landed to the right attempted to move off across the other's path to the left. Nonetheless, the 11th Manchester Regiment was able to head northwards and capture part of Kiretch Tepe Ridge, but in doing so lost touch with the rest of the brigade. Less fortunately that night, the 9th Lancashire Fusiliers thought it captured Hill 10 with great elan. Their objective turned out to be a sandbank. They rectified matters the next morning (7 August).

Until the landings got under way, 10th (Irish) Division had been 'without orders and without an objective'.[36] Now on 7 August its 30th and 31st brigades were deployed as reinforcements to strengthen the left wing of the attack. However, due to the problems at A Beach, it was

Fig. 11. 5th Battalion Royal Irish Fusiliers (10th Irish Division), Gallipoli.

deposited at two different locations—31st Brigade and two battalions of 30th Brigade were taken to C beach on the right wing instead—and the division remained separated for days, with the greater part of it far from its commander Lieutenant General Sir Bryan Mahon and its objective, Kiretch Tepe Ridge. Confusion reigned. The sun was baking hot, and water was in desperately short supply. Meanwhile, on the right flank, General Sitwell of 34th Brigade hesitated through the morning in the face of much uncertainty and a range of contradictory instructions from Major General Frederick Hammersley, commander of the 11th Division, to move forward from his Hill 10 position against Chocolate Hill (Mestan Tepe). Yet once the attack got under way in the early afternoon, it faced limited numbers of Ottomans, and the hill was taken by 5.30 p.m. Thus by the end of 7 August at Suvla, the British held the beach and the high ground immediately around it (the Kiretch Tepe Ridge as far as Jephson's Post, Hill 10, Lala Baba, Chocolate Hill, and Green Hill), and were able to make further landings of men and equipment without enemy opposition.[37] The 27,000 British and Irish troops had landed in an area defended by only 3,000 Ottomans. But they did not push forward to Ismail Oglu Tepe (W Hills) nor the Kuçuk Anafarta and Biyuk Anafarta ridges, and, rather than aiming to gain ground the next day, 8 August was largely wasted in reorganizing and consolidating. While they did so, there was time for Ahmet Fevzi's XVI Corps to march to Suvla from Saros Bay, and the 9th Division was sent to bolster Willmer's force immediately.[38]

The battles at Suvla are known in Turkish as Anafarta Muharebeler (Battle of Anafartalar).[39] Parallel discussions were taking place between the senior commanders on each side of the hill on 8 August. Fevzi, recognizing that his exhausted troops were not fit to fight after their long march in the heat, devised a plan for an attack at dawn on 9 August across a broad front from Kiretch Tepe to Chocolate Hill. He was promptly dismissed by Liman von Sanders who wanted to attack that night, but Fevzi's replacement as commander of the Anafarta Group, Kemal, agreed with Fevzi's judgement and stuck with the original plan.

Approximately 16,000 Ottoman soldiers attacked at 4 a.m.[40] On the British side, Hamilton had belatedly become so alarmed by the absence of messages from Stopford that he dispatched his staff officer, Aspinall, to assess the scene. Aspinall, deeply concerned by the lack of urgency he observed, urged Hamilton to intervene. Hamilton ordered an assault on Anafarta Ridge and set out for Suvla himself, arriving at 5.30 p.m. on 8 August. When the divisional commander, Hammersley, proposed an attack the next morning, Hamilton insisted that it must be immediate instead. However, due to the difficulty of finding and communicating with the dispersed units in the area, Hamilton succeeded only in bringing the attack forward by one hour to 4 a.m., that is, nigh on simultaneous to Kemal's attack towards them.[41] Thus, Ottoman reinforcements from the 9th and 12th divisions were in position to see off the British attack from the advantage of the high ground on Tekke Tepe and Biyuk Anafarta. Despite further British attacks on Kiretch Tepe and around Scimitar Hill over the next couple of days, including the infamous incident on 12 August where the men from the king's estate at Sandringham were lost,[42] plus a further concerted effort on 21 August, the offensive at Suvla had effectively reached stalemate and failure by 9 August. On 21 August, Stopford was replaced by Major General Henry de Lisle.

By the end of August, the extended area at Anzac and the newly won ground at Suvla were roughly joined up. A winter harbour had been secured. The cost had been terrible. Sickness had waged a terrible toll: estimates suggest that 40,000 men were evacuated from Anzac and Suvla in August as a result. The estimated British and Irish losses from the fighting at Suvla range between 5,300 and 10,000 men.[43] Overall Allied casualties in August on the peninsula (that is, killed, wounded, or missing) have been put at 45,000, of which 21,500 occurred between 6 and 10 August. Rhys Crawley suggests that, as a percentage of available infantry, at 33 per cent, the casualties arising from the August offensive are comparable to the first day of the Battle of the Somme.[44] In return, Ottoman losses for Anzac and Suvla in August were approximately 20,000.[45]

Fig. 12. A pictorial map of the Dardanelles.

Endgame

Although the Anzac breakout, in capturing two out of the three high points of the Sari Bair Ridge, came tantalizingly close to achieving its initial objective, this does not mean that victory in the campaign came close. Victory required the onward movement of these forces across the peninsula to capture the Kilid Bahr Plateau from which the Dardanelles straits can be dominated. Recent research by historian Rhys Crawley has emphatically demonstrated the impossibility of Hamilton's forces achieving this task.[46] Even without a well-organized and disciplined enemy in their way, the physical task of crossing the terrain and supplying these forces with food, water, and ammunition would have been beyond the MEF. These were sickly men who were thirsting for water: 83 per cent of Anzac losses in the first four days of the August offensive were due to sickness. The lack of water could make or break any operation, and, for example, was a further significant cause of the inaction at Suvla on 7 August. Nor did they have sufficient quality and quantity of artillery support, let alone the means to site or target them effectively. The shells crisis in May 1915 had demonstrated that Britain could not manufacture sufficient war materiel for the Western Front's requirements; Gallipoli, at the end of a six-week and 3,500-mile line of communication, fared even worse in its supplies. Altogether, this force was incapable of sustaining an assault across the peninsula.[47] If we then factor back in the considerable merits of their Ottoman opponents, the impossibility of Allied victory at Gallipoli in August is sealed. As their foremost historian, Ed Erickson argues, Liman von Sanders's forces 'reacted coherently and violently' to the renewed offensive and 'on the defensive, the Fifth Army proved to be almost an immovable object that could not be ejected from ground that it chose to hold'.[48]

The failure of the renewed offensives was clear by 10 August. Yet it was not until the turn of the year that the peninsula was evacuated in two stages. Anzac and Suvla were evacuated overnight on 19/20 December, then on 8/9 January the forces at Cape Helles followed.

One reason for the period of relative inactivity in the interim was the persistent voices within the higher command arguing for fresh attacks. The idea of four French divisions landing on the Asian shore briefly raised its head in early September, but it was inspired by the intricacies of French army politics rather than serious strategy. De Robeck's chief of staff, Commodore Roger Keyes, argued persistently for a new naval attack on the straits. Kitchener himself proposed more landings at Bulair.[49] But such schemes were gradually outweighed by more pessimistic voices. The war correspondent Ellis Ashmead-Bartlett asked visiting Australian journalist Keith Murdoch to smuggle a critical letter home to London.[50] When it was confiscated by the military police, Murdoch wrote his own version and forwarded it to the British Cabinet and his friend, Andrew Fisher, prime minister of Australia. Around the same time on 2 September, but conveying a more authoritative view, Major Guy Dawnay, Operations Division, GHQ travelled to London where he conducted a series of meetings with key decision-makers arguing for evacuation. The disgraced Stopford also submitted his own report. Hamilton was recalled on 14 October. His replacement General Sir Charles Monro spent one day assessing the situation on the peninsula, then on 31 October recommended withdrawal. Still unwilling to take this decisive step, Kitchener himself travelled to Gallipoli in early November, where he was finally persuaded of the case for evacuation.

It was a combination of the conditions and prospects on the peninsula itself and the wider strategic picture which led to the Cabinet decision of 7 December. The entry of Bulgaria into the war in October on the side of the Central Powers opened a direct rail connection from Germany to the Ottoman Empire, and with it the prospect of an inflow of war materiel, particularly artillery and ammunition, which would have been the death knell for the Allied forces. Furthermore, Bulgaria now threatened the strategic situation in the Balkans and the British government decided to switch its eastern Mediterranean resources to Salonika.

Meanwhile, a period of relative calm descended on the peninsula. Although false rumours of a massive influx of Italian soldiers prompted Ottoman reinforcements to be sent to the peninsula in September, in October three corps headquarters and with them six divisions were redeployed. Even so, there were still over 300,000 Ottoman soldiers on the peninsula and they were reinforced by the arrival of a motorized battery of Austrian 240-mm mortars and a battery of 150-mm howitzers, with 500 accompanying technical specialists in early November. Nonetheless, Liman von Sanders chose not to launch any further significant attacks, in recognition that a withdrawal was likely. Yet the timing of the Anzac/Suvla evacuation took the Ottomans by surprise. Plans were therefore developed to attack any subsequent evacuation at Cape Helles while it was under way, but only very limited exchanges eventuated.[51]

The evacuations were the only thoroughly well-planned and successfully executed Allied operations of the entire Gallipoli campaign. The manpower at the front was gradually thinned out, leaving a skeleton force in the front trenches to be evacuated at the last moment. Elaborate deception techniques were employed to maintain the illusion of normality—extra fires were lit where departed comrades should have been cooking, rifles were set up with ingenious delayed timing mechanisms, not to mention a huge mine that was detonated at the Nek. At each location, the forces left behind them some mules, and also a large pile of stores which at Suvla and Helles were torched. At 8.45 a.m. on 9 January, Liman von Sanders sent a telegram to Enver Pasha: 'God be thanked, the entire Gallipoli peninsula has been cleansed of the enemy.'[52]

Anatomy of a Victory

From the Ottoman naval victory of 18 March, via the failure to capture key points of high ground on 25 April, to the lethargic Suvla landing and the inability to hold any of the strategic points above Anzac for more than a few hours in August, the campaign was a disaster. It never secured its first-day objectives, let alone coming close to capturing the entire peninsula as a prelude to forcing the Dardanelles and intimidating the Ottoman government at Constantinople into capitulation. This was a resounding defeat for the combined forces of the British and French empires.

The French general Albert d'Amade was found wanting as a commander within his first month at Gallipoli. He was replaced by General Henri Gouraud, who was wounded and lost an arm on 30 June. General Sir Ian Hamilton's career was ruined by the campaign, and he never saw active service again. General Liman von Sanders went on to command the Ottoman forces in Palestine, where he was defeated by Allenby. He was imprisoned by the British for six months after the war on suspicion of war crimes. His post-war memoirs were translated by the US Marine Corps and studied for their lessons in amphibious warfare. Lieutenant Colonel Mustafa Kemal, who sprung to prominence during the campaign, went on to become the first president of the Republic of Turkey.

It has been estimated that 489,000 men fought in the MEF at Gallipoli.[53] Opposing them were some 550,000 Ottoman soldiers, though their numbers at any one time were never greater than 315,500 (the numbers for October 1915).[54] The Ottomans were generally outnumbered 1.5:1, but the Allied superiority in numbers was never great enough to be decisive. The Ottomans are thought to have suffered higher casualty rates (killed or missing) of 10:7.[55] One estimate suggests that there were as many as 250,000 Ottoman casualties, of whom 101,279 died.[56] The MEF's casualty figures are somewhat better documented, and have been calculated by Professor Robin Prior as 132,175 in total.[57] The figures in detail are shown in Table 1. It is interesting to note that while French and Australian losses were roughly comparable, British losses were far greater.

Yet, despite Ottoman losses outweighing those of the MEF, it is vital to note the grave difficulties in replacing Allied casualties. Hamilton was only sent dribs and drabs of reinforcements. By October, the MEF comprised 105,705 men. But if we look at how many men should have

Table 1 Mediterranean Expeditionary Force casualty figures

Contingent	Killed	Wounded	Total
British	27,054	44,721	70,775
French	8,000	15,000	23,000
Australian	7,825	17,900	25,725
New Zealand	2,445	4,752	7,197
Indian	1,682	3,796	5,478
Total	46,006	86,169	132,175

been there if each unit and formation had been full, then there would have been a further 86,453 men.[58] By contrast, Liman von Sanders enjoyed a reliable supply of replacements and reinforcements.[59] Moreover, too often Hamilton's men were inexperienced and untested in war. Nor was the MEF's leadership good enough. Hamilton himself was unduly optimistic and thereby misled his superiors in London; he and his staff devised overly complex battle plans. Too many of his subordinate officers—most glaringly at Suvla—were second rate whereas in historian Ed Erickson's summation, 'The Turks sent their best and it showed.'[60] Nor, aside from Liman von Sanders, should credit for their successes be ascribed to German leadership. By August 1915, 92 per cent of the command positions in the Ottoman Army were held by Ottoman rather than German officers.[61] Thus, the Ottoman Fifth Army comprehensively outfought the MEF: it didn't just hang on for long enough to succeed; it fought its way to victory.

5

Australia and the Civil Religion of Anzac

They waited neither for orders nor for the boats to reach the beach, but, springing out into the sea, they waded ashore and forming some sort of a rough line rushed straight on the flashes of the enemy's rifles. Their magazines were not even charged. So they just went in with cold steel, and I believe I am right in saying that the first Ottoman Turk since the last Crusade received an Anglo-Saxon bayonet in him at 5 minutes after 5 a.m. on April 25.[1]

The Anzac legend, the idea that the Anzacs were born warriors from 'a race of athletes' who were inherently brave and resourceful, was first glimpsed in this description by Ellis Ashmead-Bartlett of the landing at Anzac Cove. In a context where authoritative sources of information on the events at Gallipoli were limited, his and other journalists' despatches, and a small number of further publications, did a great deal during the war to establish the public perception of the Gallipoli campaign. What were those early sources?

The first news of the landings at Gallipoli was announced by a joint statement issued by the Admiralty and War Office on 26 April,[2] but it was not until 7 May in Britain (8 May in Australia and New Zealand) that Ashmead-Bartlett's detailed account was published. Its author was one of the British official war correspondents with the Mediterranean Expeditionary Force (MEF). An arrogant but talented writer, his vivid report was widely and prominently syndicated. Yet thereafter, as progress in the campaign faltered, he steadily became a fierce critic of its leadership. He came from the gentleman-adventurer tradition of war correspondents, and had covered many dramatic stories from around the world. It was his habit to sleep in silk pyjamas at the front, and he enjoyed throwing champagne parties while he was there.[3] At Gallipoli, he became deeply vexed by the increasingly strict censorship he faced and worried by the false optimism conveyed in General Sir Ian Hamilton's despatches. After various efforts to share his views with the British prime minister and others, in September 1915 he wrote once more to Asquith and asked the visiting Australian journalist Keith Murdoch to smuggle his letter to London. This was seized by the military police, but Murdoch wrote his own version and forwarded it to the British Cabinet and to his friend, Andrew Fisher, prime minister of Australia. Ashmead-Bartlett was sacked as a result. Shortly after, so was Hamilton. Although the commander always blamed Ashmead-Bartlett and Murdoch for his downfall, it is likely that their plot provided the occasion rather than the trigger for Hamilton's recall and the beginning of the end of the campaign.

Despite the attractive nature of Ashmead-Bartlett's prose and his exciting story, in the long run it was Charles Bean, the Australian official correspondent, who became the most influential writer on Gallipoli. Nonetheless, Ashmead-Bartlett, the master of the dramatic overview, was dismissive of Bean's meticulous approach: 'Oh, Bean,' he once reputedly said, 'I think he almost counts the bullets.'[4] Yet Bean continued his diligent work long after Ashmead-Bartlett had departed, staying with the Australian Imperial Force (AIF) to the end of the war, and then creating Australia's official history thereafter. Australian-born, but educated in Britain, Bean had forged a career with the *Sydney Morning Herald* when war broke out. He was chosen by his colleagues in the Australian Journalists' Association to be the official war correspondent with the AIF. It was agreed with the Defence Department that on his return he would write a history of the war. The expectation of undertaking these histories shaped his work as a journalist and the meticulous way in which he gathered first-hand evidence from the men. In turn, the authoritative and detailed nature of his accounts of Gallipoli has shaped all the histories that have been written in his wake.

Bean's first venture beyond reporting the campaign, however, was *The Anzac Book* which gathered drawings, poems, and stories from the men themselves. Through careful editing, Bean excluded references to boredom, futility, cowardice, or grief, and instead presented tough men who stand by their mates, using humour to endure war's hardships. Later, as planned, he edited all twelve volumes of Australia's official history, writing six of them himself over a twenty-four-year period. The first volume, *The Story of Anzac*, was published in 1921; the second Gallipoli volume came out in 1924.[5] In writing them, Bean relied heavily on his own records. There was little else to go on. As late as 1938 he explained, 'Few people realise to what an extent no guide even now exists to much of the history of the Great War.'[6] These histories therefore bear the stamp of Bean's interests and priorities, despite his conscientious efforts to achieve factual accuracy.

The most distinctive characteristic of his official history is its front-line perspective. His narrative is full of individual incidents, each supplemented by footnotes giving the personal details of its participants. This unique feature was fundamental to his commemorative aim and rooted *The Story of Anzac* in local communities and family histories. The drawback of this front-line view—as with all such books—is that it swamps the overall picture of the battle's progress, and does not linger on the strategic purpose of any given action. He may also have tended to cover up Australian weaknesses in the process.[7] But Bean wasn't trying to write for future staff officers' education, he was a journalist creating a history for a new nation. He saw his task as nothing less than 'constructing the permanent war memorial in writing'.[8] Among the questions he sought to answer was 'How did the Australian people—and the Australian character, if there is one—come through the universally recognized test of this, their first great war?'[9] Like many of his generation, his views were informed by social Darwinism and also by a deep admiration for the troops he met and interviewed. He believed that a 'race' had inherent qualities that could be generalized and that could be improved by a healthy environment. He perceived the impact of living and working in the bush as shaping a fine body of men. At the beginning of his first volume of the official history he wrote,

> The Australian came of a race whose tradition was one of independence and enterprise, and, within that race itself, from a stock more adventurous, and for the most part physically more strong, than the general run of men. By reason of open air life in the new climate, and of greater abundance of food, the people developed more fully the large frames which seem normal to Anglo-Saxons living under generous conditions. An active life, as well as the climate, rendered the body wiry and the face lean, easily lined, and thin-lipped.[10]

From these men that he admired so much, Bean thus extrapolated a national identity for Australia. This has become known as the Anzac legend.

Besides Bean and Ashmead-Bartlett, perhaps the other most-quoted author on Gallipoli has been John Masefield. His book, *Gallipoli*, was published in September 1916. It was conceived as a work of propaganda. He wrote it after a visit to Gallipoli of a few days in late September to assist a British Red Cross Society mission. With limited information available to him, instead of a history he wrote a highly romanticized sketch of the campaign. He explained in the book that he looked upon the campaign 'not as a tragedy, nor as a mistake, but as a great human effort'.[11] Without stinting the seamier details of battle, Masefield's Gallipoli is a place rich in classical associations—the Dardanelles was the ancient Hellespont, Troy is nearby—and he compares the men to the crusaders who fought the Saracens through his references to *The Song of Roland*, emphasizing their heroism and the grandeur of the enterprise. Most famously, he described the men thus: 'For physical beauty and nobility of bearing they surpassed any men I have ever seen; they walked and looked like the kings in old poems.'[12] For Britain it encapsulated a tendency to romanticize the campaign through reference to the ancient myths of the area. It was warmly reviewed in parts of Ireland and Australia,[13] but it was in New Zealand that it really seems to have struck a chord, and was often quoted in editorials there.[14]

Although censorship, purpose, and inclination meant that wartime accounts of the campaign, such as Masefield's, were broadly favourably disposed towards it, Gallipoli remained hugely controversial. Hence, the political pressure on the steadily weakening Asquith was such that in July 1916, after a series of profound political and military crises, a Royal Commission was set up to investigate what had happened. The commission gathered evidence during the following year and completed its reports in February and December 1917. Thus in a phase of the war when morale in Britain was probably at its lowest ebb, many of the country's most senior politicians and officers were called away from their war work for an inquiry which had the capacity to gravely damage imperial relations and public trust. Unsurprisingly, the resulting reports were deeply restrained. The first focused on the campaign's origin and inception and was published in March 1917, but the final report on the execution of the campaign was held back until November 1919. Its bland conclusions were connived at by extensive and secret collusion between Hamilton and a range of key witnesses.[15] The later report did not even prompt a debate in Parliament. *The Times* noted the difficulty in discerning what criticisms it actually made.[16] A gentlemanly cover-up had been achieved.

These various wartime texts—journalists' and officers' despatches from the front, works of propaganda, and official reports—along with some limited communications from the soldiers at the front, provided the basis upon which the general public of the British Empire formed its perceptions of the Gallipoli campaign. Without their response, the many thousands of words written by Bean and others might only have been petals cast on the wind that withered and died. But the public thirst for knowledge of loved ones, friends, and neighbours lent them great influence. To take just a snapshot of some of the publications in the first year or so after the campaign: in Australia, Bean's *The Anzac Book* sold more than 100,000 copies in the twelve months after its publication in May 1916,[17] and C. J. Dennis's popular story of a larrikin digger at Gallipoli, *The Moods of Ginger Mick* (published in October 1916), rapidly sold 40,000 copies.[18] In September, some of Ashmead-Bartlett's Gallipoli despatches were published,[19] and were it not for ill health, he would have completed a twenty-five-date lecture tour before Christmas (with an option of seventy-five more plus a tour of America). He was hot property: when he fulfilled his contract for twenty-five two-hour lectures in Australia in February and March 1916, he was greeted with packed audiences.[20] Ashmead-Bartlett had also taken footage on the peninsula that was released as *With the Dardanelles Expedition* at the Empire Theatre in London on 17 January 1916, and subsequently distributed in Scotland, Australia, and New Zealand.[21]

These Gallipoli texts, then, were avidly read in many different parts of the British Empire. They formed the basis of how Gallipoli was portrayed and commemorated. From shared beginnings, the remembrance of Gallipoli took very different trajectories in different countries. In Australia, there was a particularly avid interest in Gallipoli and the Anzac legend played a prominent role in Australia's developing sense of itself as a nation. But the Australian people's reverence for the Anzac legend did not remain steady; rather it waxed and waned. Its importance is such that it has been described as Australia's civil religion.[22] It will be seen that there were times when its congregation was widely predicted to die out, but instead it was born again in 1990.

A Nation in Waiting

To understand their response to Gallipoli, we need to know more about who the Australians were before Anzac. The Commonwealth of Australia was established on 1 January 1901 by the federation of six British colonies. From 1907 it was called a Dominion (along with Canada, New Zealand, and South Africa) and, while it was self-governing internally, Britain remained in charge of its foreign policy. Its ties with Britain were profound in other ways: its white population was ethnically 97 per cent British (the 1911 Census excluded full-blooded Aborigines);[23] indeed 27 per cent of the first contingent of the AIF were born in Britain.[24] Australia's currency was tied to sterling and half of its trade[25] was with Britain. Its legal system was British, its prime minister acted in the name of the Queen, and, in many respects, it looked to Britain for a lead.

Yet some elements of a distinctive Australian culture had begun to emerge, particularly in the 1880s and 1890s. Despite more than half the population living in urban areas, the vast and harsh interior of the Australian landscape shaped its sensibilities, and a characteristic bush worker, located in the outback, fiercely independent, brave, irreverent, loyal to his mates while battling the elements, was celebrated in popular literature. Meanwhile, a more egalitarian society (for whites) emerged with fair wages as a rule of thumb and the enfranchisement of women in every state by 1908, even while Aborigines were denied citizenship. Thus, it has been argued, two cultures lived side by side, an upper-middle-class British culture deeply invested in imperial patriotism contrasted with an urban and bush-derived folk culture that dreamed of a working-class egalitarian society.[26]

When the war broke out, Britain declared war on Australia's behalf and Australian men volunteered for war service in astonishing numbers. Approximately 30 per cent of males aged 15–49, around 413,000 men, volunteered during the war (conscription was never introduced), of whom 60,000 were killed or died from their wounds. Throughout the British Empire, Australia had the highest proportion of deaths among mobilized men.[27] At Gallipoli, some 25,725 became casualties.[28] In the main, the previously untried and little-trained Australian soldiers fought in the confined and rugged area above Z beach. This was initially conceived of as supplementary to the main attack at the tip of the peninsula at Cape Helles, but by August, what had become known as Anzac Cove was made the main focus of a renewed assault on the peninsula. By then, Australian soldiers had made a name for themselves as daring and fierce—if ill-disciplined—soldiers, after their assault at dawn on 25 April 1915.

Apart from a few thousand men who served in the Australian Commonwealth Horse at the end of the South African War,[29] this was the first time that men had fought in war as Australians (as opposed to as Queenslanders or Victorians in South Africa). To a culture steeped in the idea that war could be a baptism of fire that might herald the birth of a nation, the terrible losses were presented as a moment that revealed their essence.[30] Here was the genesis of an Australian identity, separate and distinct from Britishness.

This deep interest in the war, of course, stemmed from the gravity of what was at stake. The distressing news of dead and wounded loved ones had prompted rituals of grief and mourning, including religious services.[31] Early September 1915 saw first the burial of Major General Sir William Throsby Bridges, commander of the 1st Division of the AIF, near the Royal Military College at Duntroon where he had been commandant, then the unveiling in Adelaide of 'the first monument to the fallen heroes' in Australia.[32] Meanwhile, patriotic recruitment campaigns had been ongoing from the war's inception, and became ever more urgent as the war progressed, particularly as the introduction of conscription was rejected in two increasingly quarrelsome referenda in 1916 and 1917. All of these strands—ideas of war and nationhood, urgent interest and pride in events at Gallipoli, mourning and commemoration, and recruitment campaigning—came together in the development of Anzac Day.

The Origins of Anzac Day

The first ever 'Anzac Day' was a patriotic procession and carnival held on 13 October 1915 in Adelaide.[33] The carnival was an annual event that was normally called 'Eight Hours Day', but was renamed for one year and the funds raised were donated to the Wounded Soldiers' Fund.[34] It is tempting to see here, with thickly laden hindsight, the moment when the idea of Australia as the working man's paradise (the eight-hour day being a key victory in workers' rights) was overtaken by the civil religion of Anzac. The next use of the phrase was when the idea of

Fig. 13. Huge crowds attend the first ever Anzac Day commemorations in Brisbane, Queensland in 1916.

marking the first anniversary of the landings formally emerged in Brisbane in early January 1916.[35] Indeed, Brisbane, driven onwards by chaplain Lieutenant Colonel David Garland, honorary secretary of the Anzac Day Commemoration Committee (ADCC), was the trailblazer in establishing the day in Queensland, and Garland wrote to councils around Australia to encourage them to share in his mode of observing Anzac Day.[36] But the initial breadth of commemorative activities across the country was such that it must surely have been as much an organic as an orchestrated affair: led by their local civic and church leaders, many communities felt moved to mark the anniversary. It was certainly not guided from above: the acting prime minister, George Pearce, did not consider a military defeat to be suitable for commemoration.[37] Thus Anzac Day was marked across Australia in 1916.[38]

What forms did the commemorations take that first year? In the exuberant language of the sports newspaper *Referee*, 'Scenes of enthusiasm marked the celebration of Anzac Day in Sydney'.[39] Large numbers of people travelled into the city for the commemoration. At 9 a.m., traffic was brought to a standstill and there were cheers for the king, the empire, and the Anzacs. Later in the morning, 4,000 returned men marched through the city,[40] to reach the service at the Domain attended by perhaps 60,000 people.[41] In Adelaide, the traffic was also stopped at 9 a.m., but for silence. A large crowd attended the official commemoration service in Victoria Square that included stirring speeches, and there was a religious service in the town hall.[42] In Melbourne, there were religious services at St Paul's Cathedral, and 'a citizens' demonstration' in the town hall in the afternoon. Then, so as not to detract from the solemnity of Anzac day itself, buttons (that is, badges) were sold to fundraise on the following Friday (28 April). Hobart actually held Anzac Day on that Friday, which was payday, the better to raise funds for the Returned Soldiers' Club.[43] In schools across Victoria, there were educational activities prepared by the Education Department,[44] while in New South Wales many communities followed guidance issued by the premier's office.[45]

Clearly then, the first Anzac Day prompted an extraordinary response from the Australian people. It was the logical culmination of the profound interest in the war already noted, and embodied the imperatives of mourning, pride, and the pressing needs of war. These elements can be seen in Brisbane's Anzac Day. Firstly, the need for a communal act of mourning as the war unfolded was met by the religious elements in the day, wherein the clergy were able to explain the losses of war in familiar Christian terms.[46] And here, Canon Garland was particularly skilful in navigating the difficulties of drawing together all Brisbane's churches in a context where Catholics were expressly forbidden from participating in ecumenical services, lest it should confer any legitimacy upon Protestantism. Pride was expressed in the soldiers' march later in the afternoon. Finally, civic meetings were held to listen to a message from His Majesty the King, to discuss identical patriotic resolutions, and to observe one minute's silence at 9 p.m.[47] These meetings served as a recruiting tool. It was the Queensland Recruiting Committee that had organized the first meeting to discuss the possibility of Anzac Day, and the need for more recruits was universally noted, be it in the pulpit as in Brisbane or in all-out recruitment drives as in Sydney.[48]

The personnel of Brisbane's ADCC—predominantly Anglican clergy, retired army officers, conservative politicians, and businessmen—were the sort of men most likely to buy into the imperial patriotic version of Australian-ness, rather than the working-class egalitarian version.[49] But their point of view was not yet the exclusive interpretation of Anzac Day, and this can be perceived in early arguments over whether the day should be a public holiday. On that first Anzac Day in 1916 government offices, law courts, banks, and schools in Queensland closed, but the Brisbane Traders' Association declined to observe yet another holiday given the day's proximity to the Easter holidays and May Day.[50] Some of the arguments about a public holiday were couched in terms of the needs of business or the rights of the working man not to lose a day's pay,[51] but more often it came down to a battle about the appropriate solemnity of the day. Many feared that a public holiday could mean that unseemly sports, like horse racing, would mar the sacred meaning of the day. The solution in Queensland was a close holiday—akin to sabbatarianism. This was only abandoned in 1964 and thereafter hotels (i.e. pubs) and racecourses were allowed to open.[52] But elsewhere the returned men's view triumphed, particularly after they were back home en masse: it should be a day of jubilation as much as of mourning.[53] This had been the typical view among soldiers—in 1916 they'd held chariot races in Cairo on 25 April,[54] and there were always areas that favoured 'patriotic sports' on the day. In Penguin (Tasmania), for example, prearranged patriotic sports were given precedence over a memorial service in 1917.[55] What therefore emerged over time was a format for the day that resembled a funeral and a wake, with the solemnity of the church services in the morning, and in the afternoon sports but also old soldiers' camaraderie over some beer and the simple gambling game of 'two-up'. A public holiday in these circumstances meant that the maximum number of people could join in the observance of the day.

Yet as the competing interests in Anzac Day emerged, so did a strain of thought that soldiers were uniquely privileged. When a misapprehension developed that there were plans to have fireworks in Sydney on Anzac Day in 1916, soldiers slapped down the criticisms voiced by patriotic women as illegitimate.[56] Indeed women were steadily side-lined in the commemorative process. Whereas during the war bereaved mothers were accorded special status, this steadily melted away and after the war all women—bereaved or not—were merely expected to be an audience for—not participants in—commemorations.[57] They were seldom involved in organizing Anzac Day—certainly not in Queensland at least[58]—and when they dared to intrude on Melbourne's dawn service in 1938 they were fiercely criticized.[59] One reason this came to pass was the assiduous efforts of the RSL to assert the paramount status of the returned man. The RSL was a federation of veterans' organizations, formed in 1916 as the Returned Sailors' and Soldiers' Imperial League of Australia.[60] It argued persistently and successfully that war sacrifice had earned returned men a unique position in society (and special treatment from the government).[61]

Interestingly, the privileging of soldiers over women in commem-orative arrangements was unusual; indeed, precisely the opposite dynamic operated in Britain where soldiers were gradually sidelined and bereaved women given precedence.[62] There, Armistice Day was to become the main day of commemoration and the most common occasion for the unveiling of war memorials. Australia was thus also unusual in focusing these activities on the April anniversary rather than on 11 November.[63] Furthermore, the idea of making Anzac Day a public holiday was never mooted in the UK, so we can see the proposal as further indication of the importance of the day, and thereby the importance of the Anzac legend, in Australian national culture. Thus in 1923 the federal government fixed 25 April as Australia's national day and the day that Anzac Day would be observed.[64] This decision was implemented in all states by 1926.[65]

The first Anzac Day reveals the profound impact that events at Gallipoli had on the Australian people. It also shows what some people thought the campaign meant: how it was viewed, how Aus-tralian soldiers were viewed, and how Australia itself was viewed. The ideas about war and nationhood that were deeply held in the British-infused Australian culture found their fruition in the meanings that were attached to the events at Gallipoli. An editorial in the *Brisbane Courier* in the run-up to the first Anzac Day perfectly encapsulates the mixture of social Darwinist pride in race and the belief that the soldiers embodied the wider characteristics of the nation, plus the certainty that Anzacs had triumphed despite the overall defeat. It also shows that the nation's achievement and sacrifice were perfectly in keeping with that of the empire at this point in time:

> National qualities were revealed at Anzac which, apart from all military considerations, asserted the strength of race, the ability to rise superior to thoughts of personal danger, to dare to the uttermost, and to count life as nothing compared with the performance of duty. [...] The Gallipoli campaign may be written down as a series of blunders and it may be said it was a tragical failure, but Anzac was neither a failure nor a blunder. It was a triumphant demonstration of the fighting qualities and the stamina of the Australian soldiers. [...] Anzac was a national sacrifice at the altar of a great ideal—the ideal of imperial unity and the Empire's unity in the world war for freedom.[66]

Similar sentiments were expressed in editorials across the country. It was clear that, alongside pride in the undoubted bravery of the Anzacs at Gallipoli, their feats were understood at home within an imperial context. British troops were the yardsticks against which they measured themselves, even while the reality of encountering those troops had sometimes resulted in disappointment.[67] Nonetheless, the sacrifice of Australian youth was undertaken in the cause of empire.

The Evolution of Anzac Day in the 1920s and 1930s

Anzac Day had captured the public imagination from the start, and its repetition over the following years brought the sense of a gradually more established tradition. Each town made its own arrangements, with some continuing to favour day-long solemnity and others including patriotic sports and picnics after the morning's ceremonies. Across the country, villages and towns busied themselves erecting memorials to their dead; more often than not, they were unveiled on Anzac Day in the early 1920s.[68] Whilst the veneration of the dead remained, the meanings ascribed to their sacrifice began to shift. Although the tenth anniversary in 1925 brought the largest parades yet seen[69] (in states where public holidays were not yet secured, it doubtless helped that 25 April was a Saturday), the tone of the speeches and the rhetoric of newspaper editorials had changed. The purple prose celebrating the imperishable glory of great deeds for the empire was distinctly out of fashion. This, of course, had much to do with the different priorities of peacetime—the population no longer needed to be roused to further sacrifice—but there was also a distinct defensiveness to some of the comments.

In March 1925, a minor row broke out which illustrates this very issue. J. M. Drew, a minister in the Western Australia state government, directed that on Anzac Day there should be no speeches in schools that glorified war. He was reported as stating that 'the dissemination of the doctrines of militarism in the State schools at this particular epoch in our history would be very much out of harmony with the spirit of the time, when nearly all the Great Powers were endeavouring to ensure a peaceful settlement of future international differences'.[70] His decision followed a deputation from a range of groups, including several women's groups.[71] Unsurprisingly, the RSL took a dim view of the decision, and secured a change of policy. Teachers (not former soldiers, it seems) would give addresses at school on the Friday before Anzac Day, and 'these should in no way exalt militarism, but dwell on sacrifice and devotion to duty and service, as exemplified by our brave men, and our responsibility towards returned soldiers and their families, and those who fell, and upon the duty that devolved upon us of maintaining the ideals of liberty, justice and righteousness for which they fought'.[72]

The row reflects two important currents in Australian culture. First, it could be seen as another moment when the soldiers' view was respected more than that of women. Secondly, the fact that the initial decision was taken by a Labor minister may be significant. It could have reflected the cleavage in Australian politics stemming from the wartime conscription referenda, wherein the prime minister, Billy Hughes, split his Labor party and formed a new party with a 'Win the War' platform. Although Hughes lost the referenda, Labor colleagues who resisted his arguments and other anti-conscriptionists found themselves in a difficult position when it came to commemorating the war. They had been painted as traitors in wartime, and often chose to absent themselves or were excluded from the organization of Anzac Day and the erection of war memorials.[73] Anti-conscriptions had included trade unionists and many Irish Catholics, the least likely people to adopt the imperial British version of Australian-ness. It is perhaps no surprise that a Labor politician questioned the message of Anzac Day.

Some editorials on Anzac Day 1925 noted the social division. The *Sydney Morning Herald* was particularly outspoken on the matter. Anzac Day, announced its editorial,

> is a monument to patriotism. Love of country is a virtue which some seek to discredit nowadays. We hear a great deal of rather sloppy talk about internationalism, and the patriot is denounced by the doctrinaires as a 'jingo' or a 'chauvinist'. [...] Evil as was the war, it was in some measure redeemed by the noble qualities that were displayed, the courage, the steadfastness, the heroic self-sacrifice, the patriotism that blazed like a star.[74]

Here is a fierce defence of Anzac Day, but it is noteworthy that it had to be defended at all. War was no longer a backdrop to glorious deeds, but was 'evil'; and instead of the exhultation of war, came a new emphasis on the personal qualities of the Anzacs.

The early 1930s saw the completion of some of the most significant war memorials in Australia: the Shrine of Remembrance in Melbourne and the Anzac Memorial in Sydney were both unveiled in 1934. The state capitals of Tasmania, Western Australia, Queensland, and South Australia had each completed their memorial projects between 1925 and 1931.[75] Like many others, the Melbourne and Sydney memorials reflected the distinctive Australian habit of honouring both the living and the dead: an all-volunteer force prompting recognition of service and sacrifice.[76] Sydney's imposing art deco memorial in the city's Hyde Park holds a remarkable sculpture of a naked warrior lying on a shield: *The Sacrifice*. Victoria built a massive classically-inspired Shrine of Remembrance, its existence secured by the work of the ultimate living Anzac, Sir John Monash.[77] He had commanded a brigade at Gallipoli and rose to become the corps commander in France in August 1918 when the Australians played a pivotal part in the offensives that secured victory. It was not until 1941 that the Australian War Memorial (AWM) in Canberra was completed. Uniquely, it encompassed a museum of 'relics', an archive, and a 'shrine': the influence of Charles Bean in its conception is clear to see. Positioned on an axis opposite the new capital's Parliament, it is the physical embodiment of the central position of the Anzac legend in Australian national life.

The 1920s and 1930s saw not only the erection of the permanent memorials to Anzac, but the emergence of Anzac Day's most significant feature: the dawn service. The origins of the early morning gathering, which has echoes of the soldiers' wartime routine of reveille or stand-to at first light, are disputed. My research suggests that the first ever dawn service—or 'daybreak service' as it was termed—was held in Rockhampton, Queensland on 25 April 1916. It began at 6.30 a.m., and was attended by 500 people including the mayor and local church ministers. The national anthem was sung, prayers were offered, and speeches delivered.[78] Yet, although Rockhampton's Anzac Day service began at a similarly early hour in 1918, there is no evidence that it developed into a more long-standing tradition of a dawn service.[79] Three other contenders to the claim of the first-ever dawn service have been suggested: Toowoomba, Queensland in 1919, Sydney in 1927, or Albany, Western Australia in 1930.[80] In Toowoomba, one Captain Harrington led a small group of men who visited soldiers' graves at 4 a.m. on 25 April 1919 and in the following two years. In Albany, a dawn service was conducted by Reverend White, and inspired other services across Western Australia. However it appears that the earliest ceremony to catch on and establish the tradition widely was in Sydney. Some ex-soldiers on their way home from a night out in the small hours of Anzac Day 1926 came across an elderly grieving woman who was laying flowers at the cenotaph in Martin Place. They joined her in her solemn moment, and resolved to hold a service there at dawn the following year.

From small beginnings—perhaps 200 attended the third dawn service in 1929 in Sydney[81]—by 1935 over 20,000 were expected.[82] Later in the morning, assisted by free travel arrangements, between 35,000 and 45,000 were expected for the parade, as spectators and marchers. The anticipated doubling in the size of the parade meant that it was extended by thirty minutes and would be twelve, rather than eight, abreast.[83] In Perth, the parade was led by returned men, followed by ex-servicemen from elsewhere in the empire, and also by the sons and daughters of men who had died, plus members of the militia and cadets and representatives of youth groups.[84] By contrast, in Melbourne, the local branch of the RSL took a much more exclusive view—strongly deprecating the involvement of children in the parade, and of women at the dawn service. The league secretary, Mr C. W. Joyce, explained:

> It was essentially a soldiers' ceremony, recapturing the atmosphere of the vigil at dawn before the wartime 'hop-over.' The presence of other people robbed it of its significance. The League had always opposed the tendency of some returned soldiers to take children with them on the march. The children detracted from its dignity, and interfered with the step of the men.[85]

The attitude in Melbourne was thus in stark contrast to other big cities, which sought to establish Anzac Day as a possession for all the community. Melbourne's attitude was calculated to ensure the day's eventual obsolescence.

Otherwise, besides occasional continuing comments on the appropriate solemnity of the entire day, the rhetoric of Anzac Day had lost the temporarily defensive tone of 1925, as well as the worst excesses of the grandiloquent patriotism and exaggerated claims of wartime. In its place were heartfelt pride and sorrow in the heroism of Australian soldiers at Gallipoli and afterwards, to whom a debt of honour remained. As the *West Australian* argued:

> Anzac can mean nothing to us as a community and all our solemn gatherings, prayers, hymns and speeches will be only mockery and our war memorials mere whited sepulchres if the anniversary does not rekindle our desire for service towards our fellows. We are proud to say we are of the same race as the men whose deeds and deaths we ceremonially honour today.[86]

This was how the meaning of Anzac Day was made relevant to present concerns. It is important also to note that, although it was understood to have bestowed nationhood on Australia, the sacrifice of war continued to be understood in its imperial context. For example, a special issue of commemorative stamps for the twentieth anniversary of the Anzacs' landing depicted not Anzac Cove itself, or an Anzac soldier, or an Australian landmark, but the cenotaph in Whitehall, London.[87] One aspiring writer was moved to send a poem to the *Sydney Morning Herald* noting the various anniversaries that fall in 'These April Days', thereby comparing the Anzacs to the English icons St George, Shakespeare, Cromwell, and Captain Cook.[88]

Thus Anzac Day was firmly established as Australia's national day and received very active support from the Australian public. It was not subject to criticism as a concept. Indeed, it was unthinkable that it would be criticized, save for marginal issues regarding specific arrangements on the day. It was closely associated with the RSL, and hence a set of conservative and imperialist values that were hostile to class politics, and it could be uncomfortable for Labor politicians and others given the legacy of the wartime conscription referenda. The Anzac legend was couched in patriotic rhetoric about what it meant to be Australian, and it was uncontroversial to frame that discussion within the idea of the British Empire—empire as a duty, empire as a yardstick for greatness. This was to change after another world war.

Anzac Day after the Second World War

After the Second World War, the rituals of Anzac Day evolved to incorporate a new generation of veterans and the numbers of participants in the dawn service and the parade swelled: in Melbourne, as many as 44,000 people participated in the march in 1956.[89] Yet ironically, while the parades were never bigger than in this era, a period of soul-searching began. It remained mostly sotto voce while the profoundly anglophile Robert Menzies was prime minister (1949–66). The nature of Australia's role in the Second World War provided the background to these developments. The varied experiences of combat and captivity of the later war didn't necessarily fit comfortably within the straitjacket of the Anzac legend's definition of heroism.[90] The 'great betrayal' of Australia by Britain at Singapore, the ratification of the Statute of Westminster in 1942 which meant finally acquiring full legislative independence from Britain, and the strategic turn of Curtin (prime minister 1941–5) towards America, all marked significant steps away from the embrace of the mother country.[91] Later, Australia's hugely controversial involvement in Vietnam (1962–73) undermined support for Anzac Day, tainting it by association: an unpopular war caused some to see Anzac Day as an endorsement of a militarism they did not share.

Even before Vietnam, however, criticism of ex-servicemen began to emerge in public for the first time. Disagreement between the RSL and church leaders signified a partial rift in the conservative triumvirate that had inaugurated Anzac Day. Two episodes suggest the competitive tension. In 1956, there had been chaos when the march in Sydney had split to attend one of two commemorative services. The RSL had staged an alternative non-religious event to accommodate the Catholic objection to ecumenical services: perhaps it thought that the involvement of the churches was now superfluous? The result was utter confusion and the practice was dropped.[92] Then, in 1965, Anzac Day fell on a Sunday, but the RSL refused to adopt its usual practice and schedule the commemoration service for the afternoon to avoid clashing with church services.[93] Indeed, the New South Wales president of the RSL, Sir William Yeo, called for the churches to scrap *their* services to make way for Anzac Day. A Sydney clergyman, the Reverend Alan Walker, responded that the RSL was 'one of the most pagan institutions in Australia'.[94] It was as if the RSL really had come to believe that Anzac was an alternative civil religion.

The second line of criticism focused on the veterans' behaviour, which threatened to bring Anzac Day into disrepute. An afternoon of drinking had long been a feature of 25 April, but disorderliness had become a concern.[95] By 1960, a student journalist dared to describe Anzac Day as the 'annual ritual of national narcissism-cum-Bacchanalian revel', a 'protracted sentimentalising of a brutal, unlovely phenomenon', and 'conceived solely as a means for continuing the fallacy that war is an ennobling activity'.[96] The fierce response from the RSL contrasted the thousands who went to see the march with the views of 'a few cranky students'. Most famously, these critical sentiments were echoed in Alan Seymour's play, *The One Day of the Year*, which portrayed the generation gap between the disaffected student Hughie and his father, Alf, a veteran of the Second World War. The play was initially rejected for inclusion in the Adelaide Festival in 1959 for fear of offending the RSL, but was eventually performed widely around Australia.[97] It was part of the broader temper of the times that was more critical, sometimes willing to describe Gallipoli as a disaster,[98] and sometimes more suspicious of patriotism and militarism.

By the time of the fiftieth anniversary of the landings, there were concerns that the day would not survive beyond the lifespan of the veterans. The *Sydney Morning Herald*, for example, published an opinion piece that concluded that

> Perhaps the myth was sufficiently imaginative fare for the jejune Australia of the inter-war years, but the present reaction of young Australians against wartime vainglory and the political clamour by the R.S.L. suggest that the myth has outworn its utility.[99]

Meanwhile, the *Sun-Herald's* editorial argued for Anzac Day's continuance but in doing so repositioned its significance decisively away from the mourning of those with a personal connection to Australia's wartime losses towards the discovery of nationhood:

> Anzac Day remains, and should remain, our National Day not because of the losses of April 25, 1915, or of all the other days of Australia's wars, but because on that day there happened, tragically but splendidly, an event which enabled the inhabitants of six mutually suspicious London-oriented States to discover an identity—to discover that they were all Australians.
>
> More than that: it enabled them to discover [...] a classless mateship, particular kinds of gallantry, staunchness and rough tenderness—which amounted, as Dr C. E. W. Bean said, to a new or unrecognised character, an Australian character.
>
> It is in this sense that Australia became a nation fifty years ago.[100]

These comments reflect the growing emphasis on the qualities of Australian-ness that were being celebrated, and the absence of any acknowledgement of the imperial context of 1915. It is nonetheless important to note that the crowds who attended 1965's parades were still very large—100,000 people were estimated to have turned out to watch 30,000 march in Sydney.[101]

Yet criticisms of Anzac Day continued to mount. Australia brought in a form of conscription for Vietnam in 1964, and Anzac Day became a lightning rod for protest against it. On Anzac Day 1966, a group of women laid a wreath at the Second World War memorial in Melbourne as a means to express their criticism of conscription for Vietnam. As the leading historian of Australian commemoration, Ken Inglis, has noted, 'war memorials once attracted only people who accepted what they affirmed'; now that had changed.[102]

Anzac Day became the subject of other protests. From the late 1970s, small numbers of women activists started to draw attention to violence against women, specifically rape, as a feature of all wars. In some years, the police attempted to use public order powers to prevent their marches, but by 1983 had decided to take a more cooperative approach. In that year, protests took place in a number of cities across Australia, including Canberra, where 340 women marched to the AWM to lay a wreath an hour before the official march. One of their songs turned Anzac Day's liturgical rhetoric of 'Lest we forget' on its head: 'Lest they forget the countless children burned alive in napalm's fire', they sang.[103] As often as they were tolerated, such protests could provoke impassioned responses: one speaker at the AWM dawn service in 1988 identified 'radical Aborigines, greenies, feminists, and educationists' as 'the most dangerous enemies Australia faces today'.[104] Yet these protestors were always small in number. The feminist critique of war on Anzac Day is less an indication of a widespread concern in Australia with their specific arguments, and more an indication of the depleted standing of the day.

Just as corrosive as its association with militarism via Vietnam was Anzac Day's association with the imperialist conservatism of the RSL. This was a period when Australia was decisively recalibrating its relationship with Britain. Decolonization and the applications to join the European Economic Community in the 1960s marked a wholesale strategic reorientation for Britain, a process that posed significant challenges to what remained of the former Britishness of Australia (and New Zealand). Even before the symbolic changes of the Whitlam era in government (1972–5), which saw the replacement of 'God Save the Queen' with 'Advance Australia Fair' as the national anthem, and a raft of other measures, people voted with their feet on Anzac Day. It seemed a relic of the imperial past that no longer chimed with what the majority of the population felt Australia represented. Attendance appears to have reached its nadir in 1973.[105]

A Civil Religion Born Again: Anzac Day After 1990

In the post-Second World War decades, then, Anzac Day became subject to indifference, a target for criticism, and was often assumed to be dying out. However, high-profile comments suggesting declining community support for Anzac Day should be set against attendance figures for the dawn service and the morning's parade and ceremony. They tell an interesting story about changing emphases in the day's make-up and reveal a remarkable change in its fortunes.

The statistics for attendance at the commemorative services at the Australian War Memorial are incomplete and imprecise, but telling all the same. In the latter years of the Second World War, 2,000 people attended the 'main' ceremony and 'some hundreds' went to the dawn service.[106] Until 1945, the scale of the dawn service was such that it could be conducted within the cloisters of the war memorial. Thereafter it was held in the grounds as attendance expanded.[107] For the first decade after the war, attendance at the main ceremony was estimated as varying between 3,000 and 6,000 people.[108] In 1957, record figures of 1,000[109] at the dawn service and 6,000[110] later on at the main service were recorded. From the late 1950s through to the mid-1970s, the figures for the dawn service moved up and down, but crept towards 2,000, while the main service attracted around 10,000 each year.[111] By 1976, attendance at the dawn service was judged to have reached 3,000 while attendance at the main ceremony remained roughly static.[112] Attendance then continued to increase slowly, reaching 4,000 at the dawn service and 12,000 at the main service in 1984.[113] By 1991, 5,000 attended the dawn service.[114] Then a curious thing happened. The estimate for the dawn service was 10,000 in 1994;[115] 15,000 in 1995.[116] In 2005, it was 25,000[117] and 30,000 in 2008.[118] Sometime in the mid-1990s, the dawn service overtook the morning's parade and service as the most well-attended part of the day, *and* the attendance figures acquired rocket-boosters. In 1990, Australia's civil religion was born again.

How had this come to pass? Even while Anzac Day was at its most unpopular in the 1970s and early 1980s, the seeds were being sown that would serve to detoxify the elements of the Anzac legend that most grated: militarism, imperialism, and the taint of Vietnam. With dwindling numbers of ageing veterans, the stage was emptying and new stories could be told. The process is most neatly symbolized by the replacement as the archetypal Anzac of Albert Jacka VC by

Fig. 14. Thousands gather in the darkness at the Australian War Memorial in Canberra for the Anzac Day Dawn Service in 2013. The words selected by Kipling, 'Their Name Liveth For Evermore', familiar from war memorials to the fallen of the British Empire around the world are dramatically lit in the foreground.

Simpson and the Donkey. Whereas Jacka had won his VC for single-handedly killing seven Ottoman soldiers, Simpson the stretcher-bearer was a lifesaver. Here was a hero fit for a country increasingly uneasy about celebrating warfare and killing. Then the work of historians Bill Gammage and Patsy Adam-Smith, who wrote afresh about Gallipoli from the soldiers' point of view, was followed in 1981 by the hugely successful Peter Weir film *Gallipoli*, starring Mel Gibson.[119] Gammage advised the makers of the film, and Weir spoke about Charles Bean's influence on him when the film was released.[120] It duly focused on the character of the men, their devotion to their mates, and their athleticism. But in a film fit for a post-militarist and post-imperialist age, the central characters did not fire a shot, and

expressed no loyalty to the empire. Furthermore, any British characters portrayed were mocked or stereotyped as cold and aloof.

The film follows two young athletes from rural Australia, blond and
idealistic Archy (Mark Lee) and the more cynical Irish-Australian
Frank (Mel Gibson), on their journey to join up. Only a short section
at the end of the film was devoted to their time on the peninsula. The
climax of the story shows the futile attack on the Nek in August, and
amid the carnage it is reported that the British soldiers who have
landed at Suvla are sitting down on the beach to drink tea. The
contrast with the original Anzac beaching landing is implicit, and
the waste of Australian life, at the behest of what seems to be an
English officer in support of incompetent Brits, makes this a powerful
anti-war movie. In actual fact, the operation at the Nek was in support
of the New Zealanders on Chunuk Bair, and the officer who ordered
the men to certain death was Colonel Antill from New South Wales.

The film was Australian-financed and -created. Its producers were
Rupert Murdoch and Robert Stigwood, its screenwriter was the play-
wright David Williamson. It was part of a wave of Australian film-
making inspired by the new nationalism of the Whitlam era, and the
most accomplished of a number of films and miniseries, such as
Breaker Morant (1980) and *Anzacs* (1985), depicting Australian soldiers,
which served to present the Anzac legend to new audiences in the
early 1980s. Through these works, as historians Dominic Bryan and
Stuart Ward have explained, 'the notion of Australian "sacrifice" was
recast in the passive voice. Rather than willingly sacrificing their lives
for an imperial cause, Australian volunteers were depicted as having
been sacrificed by callous British officers in a futile endeavour.'[121] A neat
change of emphasis dealt with both the anti-militarist concern that
Anzac glorified war, and its imperialist association (that is, the legacy
of Britishness it represented).

In similar fashion, the former enemy has been recast as a fellow
sufferer. Although war sentiments about the enemy featured in *The
Anzac Book*, it has been shown that these were the sole creation of its
editor C. E. W. Bean.[122] There were some friendly gestures towards

Turkish comrades during the Korean War—for example, Turkish servicemen were invited to take part in Anzac Day in Australia by the federal executive of the RSL in 1953[123]—but only in the mid-1980s did such gestures gain any momentum. It was agreed in 1985 that the Turkish government would officially rename Ari Burnu 'Anzac Cove', and at the same time the Kemal Atatürk Memorial was unveiled on Anzac Parade in Canberra. It features the words from Kemal's 1934 speech which include, 'There is no difference between the Johnnies and the Mehmets to us where they lie side by side here in this country of ours'. Prior to this event, the speech does not appear to have been well known in Australia. Certainly it was not quoted in Australian newspapers around the time the words were written or for twenty,

Fig. 15. 'Those heroes that shed their blood and lost their lives…you are now lying in the soil of a friendly country': the reconciliatory words of Kemal Atatürk from 1934 are inscribed on the Kabatepe Ari Burnu Beach Memorial which was unveiled on 25 April 1985 on the occasion of the official renaming of Ari Burnu as Anzac Cove.

perhaps almost fifty, years afterwards.[124] But by the twenty-first century it was ubiquitous on official occasions and in TV documentaries, and a vital part of any pilgrimage to Gallipoli.[125] Such is its importance as the cornerstone of Australian–Turkish relations that the Turkish film-maker Tolga Örnek, who was awarded an Order of Australia Medal for his 'service to Australia' in making his documentary *Gallipoli: The Frontline* (2005),[126] was publicly rebuked by the Turkish ambassador to Australia for choosing to omit Kemal's famous speech. Thus there is now warm cooperation between Turkey and Australia in commemorating the campaign. The sympathy for the ordeal that soldiers on both sides endured has further detoxified Anzac Day.

One of the reasons why Atatürk's deeply generous and magnanimous words strike a profound chord is that Australians have always felt that Anzac Cove is sacred ground. Indeed, there is evidence of an emotional unwillingness to perceive Anzac Cove as sovereign Turkish territory, and a failure to accept that the Allied attempt to capture it ended in retreat and abandonment—a situation further complicated by the Ottoman Empire's overall defeat in the war, and the subsequent occupation of parts of Ottoman territory. In 1923, when the Imperial War Graves Commission (IWGC) was undertaking its work to establish its cemeteries on the peninsula, the Turks attempted to limit the area that might be used. This was considered to be 'impertinent' and 'scandalous'.[127] In 1953, F. C. Sillar, the IWGC secretary, further to the Treaty of Lausanne (which replaced the Treaty of Sèvres imposed on the Ottoman Empire at the end of the war), expressed his concerns about the Turkish intention to build a memorial in distinctly possessive terms.[128] Even in 2003, John Howard (prime minister 1996–2007) nominated Anzac Cove—that is, sovereign Turkish territory—as the first site for Australia's new National Heritage List. The controversies over the roadworks at Anzac in 2005 and 2007 speak to similar concerns.[129]

Bob Hawke (prime minister 1983–91) expressed most fulsomely Australian attitudes to the landscape at the dawn service at Anzac

Cove in 1990. Glossing over the fact that Gallipoli was a defeat, Hawke explained that Anzac was 'sacred' ground, and,

> because of the courage with which they fought, because of their devotion to duty and their comradeship, because of their ingenuity, their good humour and their endurance, because these hills rang with their voices and ran with their blood; this place is in one sense a part of Australia.[130]

This is part Bean, part Gettysburg Address, part Rupert Brooke ('That there's some corner of a foreign field | That is forever England'). It reverses the sentiment of Atatürk's 1934 speech, which said that the Anzacs are part of Turkey, not that Gallipoli is part of Australia: 'your sons are now lying in our bosom and are in peace. After having lost their lives on this land, they have become our sons as well.'

Hawke was the first Australian prime minister to attend the dawn service at Anzac Cove, and the elaborate arrangements made for the seventy-fifth anniversary commemoration there have been deeply influential in the Anzac revival. He had already facilitated the final aspect of the detoxification of Anzac Day and of Australia's military history through the welcome home march for Vietnam veterans in 1987—a belated acknowledgement of their service to the nation.[131] In the following year, the 1988 bicentenary celebrations uncovered profound difficulties in marking the settlement of Australia and with it the brutal displacement of the continent's Aboriginal inhabitants.[132] There was perhaps some relief that the Anzac anniversary proved more straightforward. After spending almost AUD 10 million transporting fifty-eight veterans, plus war widows, schoolchildren, ancillary staff, and seventy journalists, on a pilgrimage to the peninsula, 1990 was a huge success that marked the moment when Australia's civil religion was born again.[133]

Gallipoli's and the Anzac legend's inherent power and suitability for this central role in Australia's civil religion is further demonstrated by the contrasting approaches of Paul Keating (prime minister 1991–6) and John Howard. Keating had attempted to reorientate the nation's focus to Australia's role in the Second World War as part of a bid to position the emotional location of the country in the Asia-Pacific region rather than in the British World.[134] The reinterment of an Australian Unknown Soldier on 11 November 1993 contributed to his quest to throw off the imperial legacy. Keating also grappled with the intractable difficulties in Australia's Aboriginal relations.[135] It was perhaps with relief that John Howard once more swept aside these issues, dismissing the 'black armband' view of history, and returning to the promotion of Gallipoli and Anzac Day.[136] Thus, for example, the last original Anzac, Ted Matthews, and the last survivor of the campaign, Alec Campbell, were honoured with state funerals upon their deaths in 1997 and 2002 respectively. The sometimes challenging and unsettling Keating version of Australian-ness was superseded by a more comforting Howard version that has been implicitly endorsed by the increasing numbers attending Anzac Day.

But what is it about Gallipoli, the Anzac legend, and Anzac Day that fit so well in this role as a civil religion? Their power lies, perhaps, not just in the inherent drama and emotional pull of the events on the peninsula, but in the stable yet flexible nature of the meanings that can be attached to Gallipoli. In an Australia whose demographic make-up has been transformed through immigration, Anzac Day can be a reassuring remembrance of the country's Anglo-Celtic roots. Many people take pride in their ancestors' involvement in the campaign by participating in the march, often wearing their medals. But it has also gained an increasingly inclusive element. Demonstrating the friendly relations with Turkey, Turkish-Australians unofficially joined in the Melbourne parade from 1996 onwards, a development officially endorsed by the RSL in 2006, although the involvement of other former enemies, notably German and Japanese ex-soldiers, remained unthinkable.[137] By 2004, Turks, Vietnamese, Koreans, and Serbs all joined in Sydney's Anzac Day march.[138] Remarkably, since 2011, the Working Group for Aboriginal Rights has joined behind Canberra's Anzac Day march; insistent that this was not a protest, they co-opted the familiar language of the day: 'Lest we forget the frontier wars'.[139] The increasingly elaborate newspaper coverage of the day also actively sought out a variety of Anzac stories, including reflections of Australia's increasing ethnic diversity. Such articles include the remembrance of Aboriginal veterans in the First World War.[140] Anzac Day can thus encompass all of the nation, in a way that Australia Day never could.

However, it is the dawn service that has become the key commemorative event. The dawn service, which begins with the gathering of a crowd in darkness, has an innate drama and originality as a ritual. It is a truly Australian Invented Tradition. Its most powerful manifestation is at Anzac Cove itself. Precise figures for attendance there are hard to come by: perhaps 5,000 in 1990,[141] 10,000–15,000 in 2000, and a very overcrowded 30,000 in 2005.[142] As the numbers of backpackers attending the ceremony grew, assisted by the advent of cheap international travel, a feedback loop perhaps elevated the status of the

Fig. 16. Backpackers draped in the Australian flag and those participating in organised tours are amongst the thousands attending the Dawn Service at the Anzac Commemorative Site at Gallipoli with all its lighting rigs and other temporary infrastructure provided for the event. Attendance probably peaked in 2005 and was controlled by a ballot system for the centenary in 2015.

Fig. 17. Lone Pine Cemetery and the Memorial to the Missing which marks 4,936 Australians and New Zealanders with no known grave. Lone Pine was the scene of fierce fighting in August 1915 and is the site of the Australian national service of commemoration which follows on from the Dawn Service near Anzac Cove each 25 April.

dawn service back home. This process was reinforced by the ABC's live coverage in 1990, 2000, 2005, and every year thereafter.[143] (Dawn in Turkey is around lunchtime in Australia.) At Gallipoli, a parade was also impossible for reasons of space, so it was correspondingly downgraded. Moreover, the parade had demoted the majority of Australians to observers; by contrast, once the RSL's wish to keep the dawn service as the preserve of returned men had died away, it elevated every attendee into an equal participant. And as a ceremony which centred on a speech it had an almost infinite malleability—the dawn service is like the Whitehall cenotaph. Almost any meaning can be ascribed to the Anzacs within it.

Twenty-First-Century Anzac Day in Australia

We can trace the malleability of the meanings ascribed to Anzac Day through editorials and speeches. One of the very few constants is the idea of 'mateship', but even that can be open to interpretation. For the most part 'mateship' is used to mean the deep bonds of friendship forged under fire, as in the 2004 Anzac Day speech by the Governor General, Major General Michael Jeffery. He focused closely on military virtues in the context of battle: 'raw physical courage', 'never letting down other members of the Platoon', 'personal and group discipline', 'loyalty and pride in his Platoon', 'inspiring leadership', and 'a wonderful sense of humour and an understated deep sense of spirituality'.[144] But mateship can also be stretched in a more liberal and abstracted direction as in this 2005 newspaper comment:

> We weep for the memory of wasted young lives because in the Anzac spirit young Australians see themselves: the cosmopolitan spirit of curiosity, adventure, good humour and humility. [...] When viewed through the eyes of younger Australians, the new cosmopolitan Anzac legend still embodies the great Australian values of egalitarianism, mateship and the fair go. But these values no longer carry the Anglo-Celtic baggage they once may have had. Today, they have a universal hue.[145]

Since Hawke's visit in 1990, two other Australian prime ministers have addressed the dawn service at Anzac Cove: John Howard in 2005 and Julia Gillard in 2012. Their speeches were further evidence of the malleability of Anzac. In contrast to other Anzac Day addresses at the main ceremony, they contain no political announcements and instead are more akin to sermons in their tone. Howard's 2005 dawn service speech painted Anzac as an inspiration to present-day soldiers and civilians. With Australian soldiers once more in action around the world, he described the sacrifice of individuals, the loss endured by their families at home, and drew out these diverse Australian qualities:

99

History helps us to remember but the spirit of Anzac is greater than a debt to past deeds. It lives on in the valour and the sacrifice of young men and women that ennoble Australia in our time, in scrub in the Solomons, in the villages of Timor, in the deserts of Iraq and the coast of Nias. It lives on through a nation's easy familiarity, through Australians looking out for each other, through courage and compassion in the face of adversity.[146]

Julia Gillard (prime minister 2010–13) used her dawn service speech at Anzac Cove in 2012 to acknowledge 'a skilled enemy' for whom 'this was a defence of the soil and sanctity of home'. She noted that two-thirds of the campaign's dead came from the Turkish side. She praised their honourable generosity to their former foes:

The Turkish honoured our fallen and embraced them as their own sons. And later they did something rare in the pages of history—they named this place in honour of the vanquished as Anzac Cove. We therefore owe the Republic of Turkey a profound debt.[147]

Gillard chose not to specify the Anzacs' qualities, and said: 'In this place, they taught us to regard Australia and nowhere else as home.' Gillard, an immigrant to Australia from Wales, did not have a family heritage of service in the Australian armed forces as John Howard did. (Howard's father and grandfather had served in the AIF in the First World War.) Nor did she share his political project to promote the Anglo-Celtic aspects of Australian identity, which reached its nadir with the racist riots in Cronulla in December 2005. Rather, like Keating and her immediate predecessor Kevin Rudd (2007–10), who issued an apology to Australia's Aboriginal 'stolen generation', she, as a Labor politician, was intellectually committed to a more multicultural approach.[148] Thus in her comments we can see the completion of the ongoing moves to frame Gallipoli in terms of Australia's relationship with Turkey rather than the British Empire. She later said that Anzac Day 'represented "for all Australians" the deepest of Australian values of mateship, good humour, endurance and bravery'.[149] She went on to announce that the centenary year of 2015 would be named 'the Year of Turkey' in Australia and 'the Year of Australia' in Turkey.[150] John King, the president of the Australian Capital Territory (ACT), RSL branch, is intending to invite the Turkish military attaché to march on his right and co-lead the march on Anzac Day 2015.[151] But, as will be seen when we survey the Turkish view of Gallipoli, it is important to note that these close relations also place constraints upon what can be said about Turkey.

The naming of 2015 as 'the Year of Turkey' in Australia is just one facet of extensive plans to mark the four-year centenary of Anzac (*sic*, not the First World War) between 2014 and 2018.[152] With firm bipartisan support, Australia plans to spend twice as much as the United Kingdom on its commemorations: state and federal government expenditure has reached AUD 325 million, with a further AUD 300 million in private donations. Such figures dwarf the amount spent on the rehabilitation and care of Australian veterans. Concern for Anzac Day is not matched by knowledge or interest in Australia's Defence Force—truly Anzac Day has been divested of its militarism, but in this case not in a good way. In a devastating critique, a former soldier, James Brown has written:

> Anzac Day has morphed into a sort of military Halloween. We have Disneyfied the terrors of war like so many ghosts and goblins. It has become a day when some dress up in whatever military costume might be handy. Where military re-enactors enjoy the same status as military veterans.[153]

Bizarre projects to commemorate Gallipoli abound: there will be nine 'Camp Gallipoli' sites for families to spend time together and learn about Gallipoli (a snip at AUD 440 for a family pass),[154] an opportunity to join in an Anzac fun run ('Exercise your freedom—join the run!'),[155] or Australians can donate to the RSL by phoning a premium line to listen to a pre-recorded minute's silence.[156] Meanwhile any criticism of 'Anzackery' is howled down in below-the-line comments on newspaper websites by the voice of the common man, newly empowered in the age of the Internet.[157] Thus the Anzac revival is experiencing exponential growth in the interest and commitment to Anzac Day in the twenty-first century. Australia's commitment to Anzac Day, born again in 1990, has taken a distinctly evangelical turn.

6

New Zealand and Anzac

On 8 June 1927 it was reported that Robert Bilkey had appeared at Pukekohe Court to be prosecuted for his actions on Anzac Day that year. His alleged crime? He had been seen hoeing his turnips.

> The defendant said that farming was not his usual occupation. He was filling in a bit of spare time before he walked three miles to Pukekohe to attend an Anzac Day Service. Three of his sons had gone to the war, and one had been killed.
>
> The charge was dismissed.[1]

Two other men pleaded guilty and were fined. Since 1922 Anzac Day had been protected in law as equivalent to a Sunday. The word 'Anzac' had been protected from being cheapened through commercial use since August 1916.[2] Early twentieth-century New Zealand took the Sabbath and Anzac Day very seriously indeed.

New Zealand before Anzac Day

Like its neighbour across the Tasman Sea, New Zealand had profound links with Britain at the outbreak of the First World War. It became a British colony in 1840, and from that date until 1914 around 90 per cent of its immigrants were born in Britain and Ireland. Given the populous nature of the country, many of them were English, but a disproportionate number were Scottish or Irish and from agricultural or non-industrial backgrounds.[3] As a broad indicator of the impact of these migration patterns on the overall population, we can look at religious observance. Between 1870 and 1920, roughly 40 per cent of the population were Anglicans, over 20 per cent were Presbyterian (i.e. of Scottish extraction), 10 per cent were Methodists (i.e. usually English), and Roman Catholics (who were mostly Irish) comprised 14 per cent.[4] As a Dominion of the British Empire from 1907, New Zealand was self-governing, but Britain retained responsibility for its external relations. New Zealand used the pound sterling. Its legal system and laws were based on those of Britain. An extraordinary 81 per cent of its exports (predominantly wool, butter, and meat) and 55 per cent of its imports were exchanged with Britain in 1914.[5]

Significant levels of British immigration were preceded by Polynesian settlement dating back perhaps 500 years. The British colony was founded on 6 February 1840 through the signing of the Treaty of Waitangi, an agreement between the British Crown and more than 500 Maori chiefs. The deficiencies in its translation into Maori and the resulting differential understandings of the nature of the sovereignty that the treaty relinquished have been the subject of bitter disagreement. Indeed, land disputes caused a series of wars between Maori and Pakeha (Europeans) between the 1840s and 1872. Later, New Zealand became a global leader in progressive politics through the introduction of old age pensions, and in 1893 was the first country in the world to grant the vote to women. Its franchise had already been extended to all Maori men and all previously excluded Pakeha men by 1867 and 1879 respectively.[6]

The British declaration of war in 1914 meant New Zealand too was automatically at war. The country lent its fervent support to the cause. Approximately one in every five men in New Zealand served overseas during the war (101,000 men)—the vast majority of the total of 117,000 who served in the armed forces. Of these, 18,000 died in the war, 2,700 of them at Gallipoli.[7] An expeditionary force had seized the radio station in German Samoa on 29 August 1914, then the first contingent of 8,000 volunteers in the New Zealand Expeditionary Force (conscription was brought in from 1 August 1916) departed from New Zealand on 16 October 1914. It was formed into the New Zealand and Australian Division commanded by the British general Sir Alexander Godley, and became the second division of the Australian and New Zealand Army Corps (ANZAC). As such, it was part of the second wave of forces at Anzac Cove on 25 April, landing from about 9 a.m. onwards. Around 20 per cent of the 3,000 New Zealanders who did so became casualties that day.[8] The New Zealand Brigade of the New Zealand and Australian Division, which was to see action at Cape Helles in the Second Battle of Krithia as well as at Anzac, was subsequently joined on the peninsula by the (dismounted) New Zealand Mounted Rifles Brigade (in early May) and the New Zealand Maori Contingent (on 3 July). Fifty Maori were among the New Zealand dead at Gallipoli, from a contingent of 477 officers and men.[9]

Fig. 18. Maori Contingent, No 1 Outpost, Gallipoli, Turkey. (Read, J C : Images of the Gallipoli campaign. Ref: 1/4-058101-F. Alexander Turnbull Library, Wellington, New Zealand. http://natlib.govt.nz/records/22330949).

The pivotal incident involving the New Zealand forces at Gallipoli came during the August offensive in the Anzac sector. While British forces landed at Suvla Bay to the north, an attempt was made to break out of Anzac. Three high points—Koja Chemen Tepe, Hill Q, and Chunuk Bair—dominated this area. The Gurkhas briefly reached the summit of Hill Q before being driven back by friendly naval gunfire, and from 8 to 10 August the Wellington Battalion led by Lieutenant General William Malone held Chunuk Bair, with support from the British troops of the 7th Gloucestershires and 8th Welsh Regiment and reinforced, by further New Zealanders: the Auckland Mounted Rifles, the Otago Infantry Battalion, and Wellington Mounted Rifles. Malone was killed before a Turkish counter-attack drove his men back. The commander-in-chief, Hamilton, subsequently erroneously criticized Malone's tactics as the cause of the loss of this high ground.[10] Yet, although the New Zealanders did gain a brief, beguiling view of the Narrows ahead of them, there was no hope that the August offensive could have been successful.[11]

The New Zealanders had first heard about their force's role at Gallipoli in the same manner as Australia: through the British war correspondent Ellis Ashmead-Bartlett's breathless despatch about the landing. It was lightly doctored to include both nationalities by changing 'Australia' to 'Australasia'. The authorities in New Zealand had dithered and caused their official war correspondent, Malcolm Ross, to arrive late on the peninsula, well after the landings. But it was the penny-pinching that required him to post rather than to cable his reports home, which caused them to arrive almost too late to be useful. Like his Australian counterpart, Charles Bean, Ross was supposed to collect artefacts for a museum and to prepare himself to write the official history of the war. Ultimately, stymied by the lack of support from Colonel Alfred Robin and Minister of Defence Sir James Allen, and the absence of a Bean-like drive and vision towards a national history, Ross did not take on this role. This may be one very important reason in explaining the comparatively limited nature of the New Zealand historiography of the war, despite the work of a number of talented individuals.[12] From the outset, New Zealanders lacked distinctive and authoritative accounts of their exploits that are comparable to Bean's work.[13] How then did they perceive and commemorate the Gallipoli campaign?

The Origins of Anzac Day in New Zealand

New Zealand was inspired to mark Anzac Day by news of Brisbane's plans in late January 1916.[14] This was reinforced when Brisbane's leading light, Canon Garland, wrote to New Zealand newspapers announcing the city's plans.[15] The civic authorities took the lead in each town in organizing the day. These usually involved a patriotic event aimed at children (in addition to the patriotic lessons and speeches in school prior to the 25th), and there were a number of church services throughout the day. The main event was the afternoon's memorial service in the town hall, preceded by a parade of returned men. In Dunedin, prior to the afternoon's parade and memorial service, women from the Otago Recruiting Committee visited three cemeteries to place wreaths on the graves of war veterans and in memory of the thirteen men 'buried in strange lands'.[16] In the evening, Auckland invited 300 returned men to dinner at the town hall[17] and then staged a free evening concert of patriotic music for them,[18] later terming this event a 'smoke concert' (a type of men-only concert involving music and politics).[19] By contrast, the Wellington Patriotic Society had declined the offer of the use of a theatre along with its company and orchestra on Anzac Day since a 'vaudeville show' would be inappropriate on a memorial day.[20] Wellington preferred instead to emulate Queensland and, led by the mayor and the prime minister, held a patriotic meeting that discussed Garland's patriotic resolution,[21] with speakers urging able men to do their duty and join up.[22] There were, therefore, some minor differences in the arrangements and tone of the day across the country.

But there was no disagreement as to the day's importance. It was marked in every town and city. Some smaller towns, with their close-knit communities, saw a near-universal turnout.[23] Every newspaper produced an editorial about Anzac Day, and every church and memorial service featured numerous sermons, prayers, and speeches that were extensively reported. There are two outstanding elements in the rhetoric and tone of New Zealand's first Anzac Day. One is that in 1916 New Zealand viewed the significance of the day from an imperial not a national mindset: it saw itself as the most loyal of Britain's Dominions, and therefore did not distinguish between a national or an imperial perspective. The other—and the far more pervasive aspect of the day—is that churches, and religious or spiritual sentiment in general, figured much more prominently in New Zealand's Anzac Day than they did in Australia.

How was New Zealand's imperial mindset expressed that first Anzac Day in 1916? Speeches, sermons, and editorials emphasized that the men's heroism would never be forgotten and often encouraged others to do their duty in their stead. That duty arose from their privileges as citizens of the empire, and their ability to fulfil it reflected the influence of Britain's glorious history and the strengths of the British race. For example, in Auckland, in his Anzac Day sermon, Rabbi Goldstein said, 'How they fought to uphold British traditions can never be forgotten by us, for they live enshrined in the heart and memory of all who value England's honour.'[24] Far from the experience of Gallipoli heralding a new sense of nationhood, for New Zealanders it appears to have engendered a deeper sense of the bonds of empire. The prime minister, the Rt Hon. W. F. Massey (ironically speaking as the Easter Rising was under way in Dublin), put it in high-flown terms:

> the blood of New Zealanders, Australians, English, Irish, Scotch, and the representatives of almost every part of the Empire, was poured out in one stream. By that blood a new covenant was sealed which will knit the Empire together more strongly, more permanently and unitedly, than would be possible by any other means.[25]

With Gallipoli thus framed as an imperial endeavour, came a diligent concern to acknowledge the other parts of the empire involved in the campaign. The *Otago Daily Times* devoted the first part of its Anzac Day editorial to the 'unintentional misrepresentation' that meant the word 'Anzac' exaggerated the role of Australian and New Zealand soldiers and obscured that of other imperial forces.[26] A returned man wrote to the *Evening Post* to deplore the misnaming of the day and its inherently parochial viewpoint: the Cape Helles landing 'was an infinitely harder task than ours at Anzac'. Unfortunately, he continued, 'even in Britain a similar distorted view holds sway, [...] common justice demands that if England does not appreciate her sons' valiant efforts at their true value the colonies do'.[27] It was this sentiment that led Mr T. M. Wilford, MP to propose that Britain should be acknowledged by using the acronym 'Banzac'.[28] The idea didn't catch on.

This imperial rhetoric was part of an attempt to respond to the terrible loss of life that the war had already inflicted. The bereaved needed an opportunity to mourn their loved ones lying in graves far away, and for their loss to be given meaning. Thus the invented tradition of Anzac Day initially drew upon the established format of a military funeral;[29] its hymns, prayers, and benediction would have had a comforting familiarity and a consoling message of sacrifice and redemption. The churches, a central feature of New Zealand society, therefore had a vital role to play. The refusal of many clergymen to hold united religious services on Anzac Day was therefore particularly controversial. Catholic and Anglican priests were barred from attending such an event by their superiors. In Cambridge, the members of the commemorative committee resigned in protest.[30] In Auckland, returned men rejected the idea of a citizens' meeting to discuss a patriotic resolution and spoke forcefully of their desire for a 'service like we used to have at the front—where we all joined in together'.[31] Rotorua proceeded with a united service without Anglican or Catholic representation.[32]

The strength of religious sentiment was also reflected in a particularly intense insistence that the day was sacred: it should be observed solemnly, and sporting or commercial activity was deemed inappropriate. The Canterbury Jockey Club postponed its meet due on the day in 1916, and others followed suit.[33] As the *Evening Post* argued on the eve of Anzac Day, horse racing and betting

> would be as sacrilegious as the money-changing was in the Temple of Jerusalem. In a sense the whole of New Zealand should be as a temple to-morrow [*sic*], in honour of men who offered up their lives on the altar of patriotism.[34]

Hotels in Auckland decided to close between 10 a.m. and 5 p.m.[35] Picture theatres (i.e. cinemas) in Wellington remained closed until after the end of the memorial service at 4.30 p.m.[36] However, the Otago Employers' Association decided that its members should only close their businesses from 2.30 to 4 p.m.; that is, during the procession and other main commemorative events.[37] They were roundly condemned for their lack of patriotism as a result.[38]

It followed that there was a particularly sombre tone to Anzac Day in New Zealand, albeit with some apparent contradictions. The day's

Fig. 19. Anzac Day in New Zealand in 1919.

events were frequently referred to as 'celebrations', and businesses and ships in Wellington harbour were encouraged to display bunting.[39] Auckland and Dunedin were similarly decorated.[40] Yet, although thousands lined the streets to watch Auckland's procession of returned men, they did so, for the most part, in silence: 'Cheers there were, and applause, but an air of solemnity marked the occasion, and invested the procession with a fitting dignity.'[41] As the *Otago Daily Times* explained: 'The anniversary is no gaudy-day, to be celebrated with martial jubilation or patriotic noise. It is a solemn, we might almost say a sombre, unemotional festival, though not without a half-hidden element of pride and triumph.'[42]

This religiosity also infused the way in which the significance of Gallipoli was perceived. Commentators in New Zealand appear to have conceived of Gallipoli as a spiritual rather than a military test. This shaped their responses to criticisms of failure, and led them to emphasize the soldiers' sacrifices and their spiritual achievements rather than their heroic exploits. Thus a columnist in the *New Zealand Herald* swotted away the criticisms in the first report of the Dardanelles Commission in 1917:

> not all of the reports of the commission can alter the meaning of Anzac Day. Of course somebody blundered—that we all knew long ago [...] What we are concerned with is not the parts of this historic episode which will fade away and perish, but the parts which will live and endure, not the bungling and the muddling and the blundering but the virtues and the ideals that never die[43]

Reverend A. Liversedge of Hawera Methodist Church echoed these ideas in his 1918 Anzac Day address:

> They were not met as citizens to commemorate a great military achieve-ment, but rather to commemorate the magnificent spirit of the sons of New Zealand and glory in the possibilities of the human spirit when it was inspired by the faith of a high and worthy cause. [...] To Englishmen the story of Gallipoli may be a story of tragedy and blunder, but to the New Zealanders it is a story of triumph.[44]

Once the outcome of the war was decided, it became possible to add further claims to the Gallipoli campaign. The *New Zealand Herald's* 1919 editorial wrote, 'Always a spiritual victory, the Landing has proved, as time carries it to its proper perspective, a military victory also.'[45]

Given the insistence on the sacredness of Anzac Day—and it was not unheard of to claim it was on a par with Good Friday[46]—it was entirely consistent that it would be officially observed as if it were a Sunday. The uneven arrangements for public holidays and business closures on the 25th had been an issue from the start. Any suggestion that Anzac Day might be marked on the nearest Sunday was liable to strong criticism.[47] In 1920, the New Zealand Parliament passed a law establishing the day as a holiday, with no race meetings to be held and licensed premises to be closed.[48] This was three years ahead of the beginning of similar formal moves by the federal government in Australia. In 1922 the New Zealand Parliament clarified the law so that Anzac Day should be treated as a Sunday regardless of which day of the week it actually fell.[49]

Before that, the fervently religious atmosphere in the early 1920s was such that an outspoken columnist could describe the opening of shops on the 25th as a sacred day 'befouled by sordid money-making, degraded by greed, and made a mockery by callous indifference to the sacrifice of the men'.[50] That columnist wrote on behalf of the Returned Soldiers' Association (RSA), and returned men were instrumental in tightening up the observance of the day once they were back in numbers at the end of the war. Already in 1918 the Returned Soldiers' Conference had resolved that Anzac Day should be observed as 'a close holiday, similar to Good Friday in every respect', a day of national mourning in which amusements were inappropriate.[51] Even so, some quarters of New Zealand society saw the first post-war Anzac Day as an opportunity for sports and dances[52] (including some RSA branches), but others responded with consternation. The *Ashburton Guardian's* Anzac Day 1919 editorial enquired,

Were the soldiers butchered to make a dancers' holiday, or to give banks another excuse for closing their doors? Are picnics the form of wreaths we are to place, as it were, on the graves too far off to be visited. Are feastings or fastings to prevail?[53]

The result was that throughout most of New Zealand 'any disposition to make high festival of the occasion had been severely crushed in its inception', and the RSA had led the way in this.[54]

By 1920 the RSA's membership had reached its pre-Second World War peak at 57,000 members, equivalent to two-thirds or more of all returned men.[55] Just as in Australia, the return home of ex-servicemen had seen them use their influence to shape the tone of Anzac Day. But whereas in Australia, returned men had encouraged an element of celebration, in New Zealand they had helped to crush it. In contrast to their behaviour while overseas, the returned soldiers' attitude had become much more serious once they came home. In a rare example of condemnation, a satirical column in 1919 criticized the result:

New Zealand lacks the real festival spirit. [...] In our opinion there can't be 'too much sport' but there can be too much gloom, too much hypocritical and solemn humbug, too much measured gravity, too much wowserism [i.e. being a killjoy].[56]

Anzac Day in the 1920s and 1930s in New Zealand

By the tenth anniversary of the landings at Gallipoli, Anzac Day was very firmly established in New Zealand. Since 1922, it had been preceded by 'Poppy Day', a fundraising day for the RSA—modelled on the British, Canadian, and Australian models. Uniquely, New Zealand sold its commemorative poppies around Anzac Day rather than Armistice Day as a result of a delay in the first shipment of poppies that had been due in late 1921.[57] Anzac Day was Pakeha-dominated, but not to the exclusion of Maori, and Maori-specific First World War memorials were erected in this period.[58] It has been argued that Anzac Day's existence and importance are proof of a developing national sentiment in New Zealand.[59] Yet, although in both countries this sentiment was entirely consistent with imperial patriotism, in comparison to Australia, New Zealand's new national pride seems more limited and understated. There was no grand nation-building project, akin to Charles Bean and his ambitious official history or to the construction of the Australian War Memorial. The plans for a government-funded memorial in New Zealand lapsed during the 1920s, and only came to fruition through private subscriptions (along with government subsidies) in 1932.[60] The nationalist sentiment that has been identified was submerged once more in the conservative political atmosphere of the 1920s.[61] Instead of the emerging Australian emphasis on the personal qualities or achievements of the Anzac soldiers, the editorials and sermons on Anzac Day 1925 in New Zealand tended to emphasize the inspiration of the soldiers' sacrifices. Here is a relatively rare (and rather modest) reflection on those qualities from the *Northern Advocate*: 'at Anzac the mettle of the Dominion's manhood was proved and a standard of courage and battle discipline was set that lasted throughout the war'.[62] Breathless commentary on their exploits in battle was notably absent. Instead abstract statements regarding the overall aims and achievements of the war were preferred, such as this from the *Auckland Star*:

> our losses were worthwhile. We were not fighting for material gain, nor for any increase to our lands, but we were fighting for honour, liberty, and the ideals of our race. It is this which gives the real meaning to Anzac Day. For on that day we proved that we were one with all other parts of our Empire in courage, sacrifice, and devotion to the parent land.[63]

Reaffirmations such as this of New Zealand's part in the empire and the empire's role in the world remained commonplace. Finally, the inspiration of the Anzacs' sacrifice to present-day action was noted repeatedly: 'the call of Anzac is the call to service'[64] as Reverend T. Fielden Taylor put it at Wellington Town Hall. Hence New Zealanders, it was argued, should heed the example of unselfishness and brotherhood to work together: 'Anzac Day reminds us that individuals must not live to themselves alone, but must recognise the claims of citizenship and nationhood.'[65] Such opaque effusions were the only hints of disunity in New Zealand in 1925, and the defensive tone of Australian editorials of the same year was entirely lacking.

Thus the first ten Anzac Days in New Zealand saw only minimal changes in the tone or the nature of the arrangements, particularly as most towns adopted the arrangements for a parade and military burial service set out in 1920 by RSA president Dr E. Boxer.[66] Yet the public gradually assumed a greater role in the services from the early 1920s, as the erection of war memorials provided a new location for the services. Freed from the constraints of space in the local town hall, the general public became more actively involved and the laying of wreaths by bereaved relatives became a vital and deeply moving part of the ceremony. Indeed, the laying of wreaths at war memorials listing the names of the dead became a distinct focus in New Zealand, compared to the Australian emphasis on the parade to honour its volunteer veterans. In the same process, and with the ritual of Anzac Day more firmly established, the need to borrow from religious discourse declined[67] and, with it, so did the role of the clergy. Increasingly, it was local politicians or returned servicemen who made speeches at the ceremonies.[68] During the 1920s, the focus also shifted from the needs of bereaved relatives towards attempting to inspire a younger generation.[69]

It follows then that there gradually emerged in the 1930s a greater willingness to query the solemnity of the day, prompted by questioning and comparisons from abroad. The (British) Governor General, Sir Charles Fergusson, led the way in 1929 suggesting that Anzac Day was too mournful and too lengthy. He hinted that some had found the day an ordeal:

> We ought to put sadness out of it and have more glory introduced. [...] We do not want it to be a hard, sad day, but a ceremony that is more cheerful—one which all will look forward to, and in which all will delight to take part.[70]

Change came slowly. The Funeral March was omitted from Welling-ton's Anzac Day service from 1934 onwards. The local RSA executive took its lead in this from the Anzac Day arrangements in Sydney, its chairman noting, 'The keynote of the ceremony is surely not one of gloom, but rather of triumph.'[71] In 1936 the funeral of George V was marked by two minutes' silence rather than a day of mourning in Britain, and in that context, Anzac Day in New Zealand began to look out of step.[72] Nonetheless, the decision to open the cinema in Dun-edin during the evening of Anzac Day that year sparked a flurry of discussion. Some correspondents to the newspaper on the subject acknowledged that thousands of people, the younger generation in particular, took the day's holiday as an opportunity for a picnic.[73] Yet the RSA as a body continued to insist on its strict observance.[74]

Interestingly then, it was the experience of a 1,400-strong party of New Zealand returned men who observed Anzac Day in Sydney in 1938 that was influential in the most important change before the Second World War. They reported back: 'We had formerly held the opinion in New Zealand that the Australian observance lacked solemnity, but now our impression is that the Australians treat the occasion at least as reverently, but more rationally.'[75] They were impressed by the dawn service and particularly valued the oppor-tunity to 'freely fraternize' after the morning's ceremonies. The first dawn commemoration in New Zealand appears to have been a dawn parade of returned men, held in Wanganui in 1935.[76] It was also discussed in Dunedin[77] and in Auckland in 1938, directly inspired by the simple soldiers' ceremonies held in Australia. Auckland went ahead with the idea, and the Governor General and the prime minister agreed to attend the inaugural event in 1939. Another was held in Wellington. The Auckland dawn parade was organized by a newly formed Australian Imperial Force (AIF) Ex-Servicemen's Asso-ciation.[78] Indeed, Anzac Day, and particularly the dawn service, can be viewed as Australia's greatest export. In some regards, it was an ahistorical development for New Zealand, as the *Auckland Star* dutifully noted:

It is pointed out that, historically, the New Zealanders do not hold a dawn parade, as the men from the Dominion did not land on the Peninsula till some hours later on the morning of April 25, 1915, though the Australian forces went ashore at the break of day.[79]

This is typical of the self-effacing nature of New Zealand culture, though it should be noted that the dawn service is not just an echo of the timing of the landings, but also of the morning 'stand-to' on the Western Front. The development of what was to become its most important national day was distinctive to New Zealand in its solemnity and religiosity. Yet Anzac Day always drew meaning from New Zealand's role in the family of empire, and its very existence, and what was to become its key characteristic, the dawn service, were not home-grown but instead inspired by Australia.

Anzac Day in New Zealand after the Second World War

As another war broke out, Anzac Day continued to be marked throughout New Zealand, often featuring exhortations to live up to the example of the first Anzacs through service and sacrifice in this Second World War.[80] The only exception was the cancellation of the dawn service in 1942, seemingly for fear of aerial bombardment in the wake of the fall of Singapore (the reason given was the 'exposed position' and 'direct light from torches and vehicles').[81] Australia cancelled Anzac Day completely that year.[82] Thereafter, in peacetime, events took a seemingly contradictory course. At the end of the war, attendance at commemorative services was boosted significantly, and Anzac Day's protection in law was reiterated in 1949. Its meaning was also officially expanded to encompass the commemoration of the Second as well as the First World War, plus the Boer War. But then, as the historian Helen Robinson has shown, in the late 1940s attendance dropped sharply and Remembrance Sunday suffered 'an immediate collapse': estimated numbers at Auckland's Anzac Day citizen service were 30,000 in 1946 but only 5,000 in 1948. She suggests that the key reason for the decline in attendance was the reduced need to mourn the dead, compared to the shocking losses of the First World War.[83] But it perhaps also reflects the declining relevance of a day that elevated Gallipoli veterans above all others; as a result, veterans of the Second World War viewed the day with far less reverence than the previous generation.[84]

Attendance remained relatively steady at the new lower rate. In 1955, the *Auckland Star* noted a 'record dawn parade' of perhaps 3,000 people, while 6,000 watched 1,957 men and women marching to the Cenotaph in the afternoon.[85] In this era, friendly relations with erstwhile foes developed, even while attitudes towards Britain became more critical.[86] By 1965, any notion of the inspirational sacrifice ascribed to Gallipoli in the pre-1945 era, was long gone to be replaced by brisk assessments of the campaign's military history. The *New Zealand Herald*'s editorial for the fiftieth anniversary, for example, dwelt on what was now regarded as its strategic brilliance, but also its 'fruitless valour, bitter defeat and unexpected victory': it had been a 'sombre pilgrimage for the Anzac veterans' including the New Zealanders who had returned to the peninsula for the occasion.[87]

As in Australia, criticism of the day developed in the post-war era. In New Zealand, the immediate focus was on the straitjacket of gloominess imposed by the requirement for the day to be observed as a Sunday. The double standard wherein some returned men enjoyed a boozy afternoon, while the RSA campaigned to halt leisure activities for the general public drew particular criticism.[88] After years of discussion, in 1966 the afternoon of Anzac Day was 'Saturdayized' (to use the local vernacular); that is, sports and leisure activities were permitted. Meanwhile, like Australia, New Zealand had become embroiled in the Vietnam War alongside the United States of America, but not Britain. After sending first surgeons in 1963, then engineers in 1964, New Zealand finally sent combat forces: a field artillery battery in 1965, followed by two infantry companies in 1967, and later small numbers of naval, airforce, medical, and training personnel.[89]

Involvement in the war was controversial, although the absence of conscription in New Zealand prevented the anti-war movement from becoming truly broad-based. Anzac Day became a focus for protest between 1967 and 1972 including the laying of wreaths carrying anti-war slogans, marches, and placards.[90] It has been claimed that these protests marked the end of the RSA's control of Anzac Day; certainly they were an attack on its reverence for sacrifice in war.[91]

The post-war decades were a period in which New Zealand's national identity was in flux. The country's close ties with Britain were profoundly shaken by the application to join the European Economic Community (EEC), which heralded the end of unfettered access to Britain for New Zealand's goods and peoples. It had become a foreign country. (Later, the immigration patterns of the 1980s and 1990s steadily made New Zealand more ethnically diverse and less British.) But it was Waitangi Day rather than Anzac Day that was the main focus for debates about New Zealand's identity: some people advocated the cancellation of Anzac Day as a public holiday in favour of Waitangi Day in the early 1970s. Yet, the replacement of a complacent idea of the country's racial equality with a greater understanding of its tensions and difficulties, made Waitangi Day a vexed occasion. Where Maori veterans spoke at Anzac Day, a similarly changing pattern in their rhetoric may be discerned: the 1950s' emphasis on loyalty and unity was replaced over time by calls for shared sacrifice to be matched by shared reward in peacetime.[92]

Anzac Day attendance was in the doldrums in the 1970s, but the seeds of its revived fortunes thereafter were sown in a similar manner to Australia: through the impact of historical scholarship and the retelling of the story for a popular audience. Building on the growth of an academic infrastructure and a popular market to support more diverse and questioning studies of New Zealand history from the late 1960s,[93] came the 1980s' investigations of New Zealand soldiers' experiences in the war by Chris Pugsley, Jock Phillips, Nicholas Boyack, and others.[94] Pugsley's book[95] marked the first scholarly study of the New Zealanders at Gallipoli: a bold first step towards putting the NZ back into Anzac in a historiographical context dominated by Australian accounts. In 1982 the novelist Maurice Shadbolt drew on Pugsley's work and his own oral history book to produce a play, *Once on Chunuk Bair*, which aimed to rehabilitate the reputation of Malone and the Wellingtons. Through it he asserted the pivotal nature of the August offensive at Gallipoli and of the capture of Chunuk Bair. Indeed, in his oral history, *Voices of Gallipoli*, he went as far as to argue (implausibly):

> Viewed from the twentieth century, the implications of success seem stunning. They suggest that we could have been inhabiting a vastly different planet—without a punitive Treaty of Versailles, and Adolf Hitler; without a Bolshevik Russia, and a Joseph Stalin; without Auschwitz, the Gulag Archipelago, Hiroshima and a hundred other horrors. The taking of Chunuk Bair; it is possible to argue, might well have given the twentieth century a last chance to be otherwise. At least it can be said New Zealand has never been nearer deciding the destiny of great nations.[96]

The play also presented the campaign with a distinct anti-British tone: not only had the British failure at Suvla played its part in the loss of Chunuk Bair, but Malone—a talented and independent-minded officer who protected his men from suicidal orders—had been killed by a British shell. The play was not immediately popular,[97] and, despite its distinct parallels with Peter Weir's *Gallipoli*, it did not reach a wider audience due to the disappointing nature of the 1991 film version. It has not therefore had the same impact as Weir's blockbuster.[98]

The play was intended as a cipher for current events: the British had failed Malone and the New Zealanders at Chunuk Bair, just as they had later betrayed New Zealand by joining the EEC. It was produced at a moment when New Zealand was receptive to a more independently nationalist stance. In 1985, the adoption of a 'nuclear free' policy by David Lange's government (prime minister 1984–9), protected in law from 1987, heralded a fierce row with the United States of America and the end of New Zealand's participation in the Australia, New Zealand,

Fig. 20. Two competing versions of the significance of Chunuk Bair. In the foreground is the New Zealand battlefield memorial honouring soldiers 'from the uttermost ends of the earth', that was unveiled in 1925. It is the site of the New Zealand Anzac Day ceremony each year. In the background is the slightly larger Conkbayiri Atatürk Memorial, erected in the early 1990s. This statue of Mustafa Kemal, later first President of the Republic of Turkey, who was instrumental in dislodging New Zealand and British soldiers from their short-lived position on the high ground of Chunuk Bair in early August 1915.

and United States (ANZUS) Security Treaty that had previously propelled its participation in Vietnam.[99] France's 'state terrorism' in sinking the Greenpeace flagship *Rainbow Warrior* in Auckland harbour, and the outraged response to it, further cemented a growing sense that New Zealand would forge its own moral path in the South Pacific.[100] Anti-nuclear activists laid wreaths on Anzac Day during the 1980s, as did feminists, gay rights campaigners, and Maori activists: the day had become a lightning rod for comments on war and society in

New Zealand.[101] An Anzac Day editorial on the seventieth anniversary used the war and nuclear weapons as emblems of New Zealand's journey towards independence:

> The young men of New Zealand and Turkey slaughtered each other in defence of their respective empires. For both countries the battles at Gallipoli marked important stages in their growth as independent nations. [...] The seeds of New Zealand's decolonization process sowed at Gallipoli took a long time growing. We were reluctant to change from 'colony' to 'Dominion' when it was offered by the Imperial Parliament and few disagreed when we leapt to Britain's defence in 1939.
>
> In recent years, Britain and West Europe's decline as major world powers has been mirrored by New Zealand and Australia's growing realization that they are South Pacific nations with close Asian links. That our interests no longer automatically coincide with those of Europe.
>
> Thirty years ago, Australia allowed Britain to detonate deadly, experimental nuclear weapons in its deserts. The concept is unthinkable now.[102]

Anzac Day in New Zealand since 1990

With a developing nationalist sense and accessible New Zealand-focused versions of the events at Gallipoli now available, the stage was set for a growing interest in the campaign in the late 1990s. This Anzac revival was more modest than Australia's. Popular and political support for the promotion of Anzac Day came later. The growth in New Zealanders' attendance at Anzac Day services at home and abroad has been widely noted—accepting, of course, that it is impossible to disentangle the numbers of New Zealand from Australian backpackers at Anzac Cove. It is probable that the Australians led the way. Whereas in 1990 the attendance of the Australian prime minister Bob Hawke and his emotional speech at Anzac Cove were pivotal in the renewal of his country's commitment to commemorating Gallipoli, New Zealand sent its Governor General, Sir Paul Reeves.[103] Later in the decade, a delegation led by the New Zealand Speaker of the House of Representatives visited Turkey, and included Gallipoli in its itinerary. In its wake, a report written by Philip Burdon, trade negotiations minister, recommended an increase in funding for commemoration and greater ministerial representation at such events.[104] It was only in 2000 that a New Zealand prime minister attended the dawn service at Gallipoli.

Helen Clark (prime minister, 1999–2008) shared with John Howard, her Australian contemporary, a personal link to the Gallipoli campaign. Among her ten great-uncles who served in the First World War, one had died on Hill 60 in August 1915. She had already visited the peninsula in 1995, and was knowledgeable about her family's history. She was simultaneously the minister of arts, culture and heritage, and she took on that role to encourage the expression of the nation's identity through a 73 per-cent increase in funding for culture between 1990/1 and 2003/4. The interment of a New Zealand Unknown Warrior in Wellington in 2004 was only the most prominent of many initiatives to promote the country's military history, including new war memorials around the world, prime ministerial attendance at commemorative services overseas, essay competitions, forewords written by Clark, book launches at Parliament, and in 2005 a plaque to honour Malone. The funeral procession preceding the interment was attended by perhaps 100,000 people.[105]

Clark's speeches at Gallipoli or at the interment of the Unknown Warrior set out her reasons why it was important to remember the First World War. They usually began with a Maori greeting, and they always noted the sacrifice and loss of war, and its impact on all families and communities. Clark's speeches had little in common with the soaring rhetorical style of Hawke in 1990 or Keating's oratory at the interment of the Unknown Australian Soldier. But in their own prosaic way, they set out a heartfelt case for New Zealand's identity. Clark asserted at the dawn service in 2005 that 'It was here that our young nations began to become of age. It was from here that we began to think of ourselves as not just servants of the British Empire, but as distinct national entities.'[106] The original New Zealand emphasis on the country's devoted role within the empire was virtually absent; indeed Britain's share in the sacrifice (and India's and France's) was noted in her 2000 speech but not subsequently.[107] Instead, she developed the themes of brotherhood with the Australians and reconciliation with the Turks, through Atatürk's words: 'You the mothers who sent their sons from far away countries [...] they have become our sons as well.'[108] Yet there was no attempt to set out distinctive New Zealand qualities akin to the Australian habit of expanding on the Anzacs' mateship, courage, and good humour. Rather, Clark preferred a vision of what New Zealand stood for in the world: that the service of individual New Zealanders through its military engagements overseas during the century since Gallipoli including in peacekeeping operations, had helped to shape the world we all live in. This sentiment concluded her speech at the interment of the Unknown Warrior:

> The Tomb of the Unknown Warrior symbolises the very personal tragedies New Zealanders have endured in our engagement with wars overseas. Let it also be a testimony of pride in New Zealand's contribution; a reminder of the heroism of our people; and a symbol of our ongoing commitment to a more harmonious and peaceful world.[109]

New Zealand has allocated more than NZD 19 million to fund its First World War Centenary programme for 2014–18. Its objectives retain Clark's vision of the role of commemoration in preserving New Zealand's military heritage, exploring its national identity, and strengthening its foreign relations.[110] The centenary of Anzac Day will be the highlight among the coordinated commemorative and educational projects of this programme. Its position as New Zealand's national day has been secured by the revival of interest in it in the twenty-first century. Anzac Day's ever-growing prominence and support from the public means it has experienced a reversal of fortunes and now overshadows Waitangi Day (which, borrowing from Anzac Day, has lately been marked by its own dawn service).[111] But this is no rerun of Australia's divisions over Australia Day, Aboriginal rights, and black armband history, for, although Maori–Pakeha relations are not without their difficulties, the emergent biculturalism and multiculturalism of the late twentieth century are indicative of a more respectful and thoroughgoing accommodation between the two cultures.

In August 1915, as one member of the Maori Contingent wrote home, 'every trench we took, you'd hear the cheer before charging, and the Maori war-cry'.[112] Ninety-five years later, it was heard again on the peninsula. Following in the footsteps of a growing number of youthful New Zealand backpackers, in 2010 a New Zealand surf club travelled to Gallipoli to take part in a sports tour that embodied national friendship and reconciliation between the people of New Zealand, Australia, and Turkey. The New Zealanders' performance of the haka at Chunuk Bair was a sincere and moving expression of a bicultural national identity at what has become a sacred site in the birth of a distinct New Zealand identity.[113] New Zealand has moved decisively beyond its affinity with the British Isles, to develop its own distinctive identity as a world citizen from the South Pacific. Yet, ironically, one of the key vehicles to achieve this has been its unique version of Anzac Day in commemoration of its service and sacrifice at Gallipoli as part of the British Empire.

7

Britain and Ireland: Gallipoli Day or Anzac Day?

O n 18 September 1933, the owner of a travelling cinema called William Walsh was held up at gunpoint in the town of Gort in Galway. The gang of armed men, members of the Irish Republican Army (IRA), stole his copy of the film *Gallipoli*. At the men's trial for larceny there were cries of 'Up the Republic' and 'Down with the British'.[1] Walsh, who had been showing the film all over the Irish Free State since the beginning of the year, had been warned by members of the IRA earlier in the month not to show the film because of its portrayal of British troops in action.[2] The film was later deemed to be 'British propaganda' by the County Galway Executive of the IRA.[3] Yet the film had been showing in Dublin to 'crowded houses' since 1931.[4] Based on Ernest Raymond's bestselling novel *Tell England* and known in Britain by the novel's title, it was directed by Anthony Asquith (son of the former British prime minister) and Geoffrey Barkas.[5] *The Times* later judged it to have 'scored a triumphant success'.[6] The stark contrast between its reception in Britain and in some quarters in Ireland illustrates how differently the Gallipoli campaign might be perceived in John Bull's two islands.

A Very British Campaign

Although the Gallipoli campaign is best known for the involvement of Australian and New Zealand soldiers, the largest single contingent within General Sir Ian Hamilton's Mediterranean Expeditionary Force (MEF) actually came from the British Isles. They were given the most important tasks at the outset of the campaign, and they bore by far the largest number of MEF casualties. The greatest share of the responsibility for the campaign's failure lies with the British. It was the War Council of the imperial government in London that decided on the strategy. It was these same men who made the fatal errors in delaying a joint navy–army attack. They also starved the campaign of resources, ensuring its failure. Furthermore, the most important commanders in theatre were from the British Isles. Thus, although men from around the world fought this campaign, its failure was above all British.

The professional soldiers of the 29th Division provided the main thrust of the invading force which landed at the toe of the peninsula at Cape Helles on 25 April 1915. Among them were the Welshmen of the South Wales Borderers landing at S beach, the Irishmen of the Royal Dublin Fusiliers and the Royal Munster Fusiliers landing at V beach, plus men from the north of Ireland in the Royal Inniskilling Fusiliers landing at X beach, the Englishmen of the Lancashire Fusiliers landing at W beach, and the Scotsmen of the King's Own Scottish Borderers (KOSB) landing at Y beach. Later the Territorial Force joined them, as did New Army divisions recruited at the outbreak of the war, including the 17,000-strong 10th Division mostly raised from across Ireland. Altogether, there was scarcely a corner of the British Isles that did not send men to Gallipoli: from the Highlands of Scotland and rugged countryside of Northumberland to the flat expanses of Lincolnshire and East Anglia or the green fields of Ireland, from the men of the king's estate at Sandringham in Norfolk and the mounted yeomen of Berkshire to the industrial heartlands of Glasgow and Lancashire.

In 1915 all of these men were citizens of the United Kingdom. By 1922, the Irish Free State was a separate country. While these men all fought at Gallipoli in 1915–16 as members of the British armed forces, the way in which the campaign and their contributions to it were perceived and remembered took radically different trajectories. This was a moment when the nature of Britishness was in flux. An all-encompassing imperial identity began to atomize. At the outbreak of the war it was possible to consider oneself to be simultaneously British and Australian or British and Irish, but those identities were later rent asunder. In Australia it happened slowly and peacefully; in Ireland it happened rapidly and violently.

In Ireland, long-standing tensions in its relationship with Great Britain came to the boil between 1916 and 1921. As the war broke out, the historic goal of achieving Home Rule was on the verge of attainment. To help seal the deal, the Irish nationalist leader John Redmond first assured Parliament that Ireland would be loyal during the war, and then once the Irish Home Rule Bill became law in September he encouraged Irishmen to join the colours and fight abroad. Many of them joined the 10th (Irish) Division, the first-ever specifically Irish higher formation in the British Army. Their opening taste of battle was at Suvla in August 1915, the most disastrous episode of the campaign. Meanwhile, during the war, Redmond's moderate stance was outflanked by more radical nationalists, who seized the opportunity of war to rise up against the British Empire and rid Ireland of the British. Their attempted armed uprising at Easter 1916 began on 24 April, the eve of the first anniversary of the landings. It initially met with minority support, but as the British brutally botched their response, Irish opinion hardened and swung in favour of outright independence. This was achieved in 1922 after a vicious guerrilla war. As the fledgling nationalist government worked to establish the Irish Free State there was an understandable atmosphere of Anglophobia, and the sacrifice and slaughter of Irishmen in the service of the Crown was not a memory they cared to burnish. Hence the IRA's highway robbery in 1933, as well as countless other acts which suppressed the public commemoration of Gallipoli and of the First World War more generally.[7] It was left to those Irishmen who did wish to mark their participation to do so unobtrusively. It was only in the 1990s that this situation changed.

Gallipoli Day or Anzac Day?

The Britishness of the Gallipoli defeat notwithstanding, to write about the memory of the Gallipoli campaign in Great Britain and Ireland is, in large measure, to write about the commemoration of the Anzacs rather than the more numerous participants from the British Isles. On the first anniversary of the invasion at Gallipoli, in a scheme most likely dreamed up by the visiting Australian prime minister Billy Hughes and the journalist Keith Murdoch, 2,000 Australian and New Zealand troops marched along streets lined by thousands of people through London to Westminster Abbey. The memorial service there was attended by the king and queen, the archbishop of Canterbury, Lord Kitchener, Hughes himself, General Sir Ian Hamilton, and the British commander of the Anzacs, Lieutenant General Sir William Birdwood. *The Times* dubbed it 'Anzac Day'. A wreath was laid by the New Zealanders in honour of their brothers-in-arms, the 29th Division. The event had been promoted assiduously in the newspapers beforehand, and was a propaganda coup. According to *The Times*, the day provoked patriotic fervour and flag-waving as had not been seen since the outbreak of the war.[8]

The nature of the patriotism echoes the contemporary view from elsewhere in the empire. It was simultaneously national and imperial: the men had died for king and empire, and had in doing so cemented the existence of a brotherhood of British peoples across the empire. At this stage in the war these imperial bonds were in no way incompatible with a heightened sense of Australian identity. As Billy Hughes told convalescing Australian soldiers prior to Anzac Day, 'Always we have been prouder to call ourselves Australian since you made our name for us in Gallipoli.'[9] Indeed, in the *Manchester Guardian*'s view another apparent indicator of a trajectory towards separate identity, that of self-government (that is, Federation in 1901), was the crucial precursor which

Fig. 21. Large crowds watch Australian and New Zealand soldiers march past on the Strand for the first Anzac Day in London 1916. Anzacs dominated the commemoration of Gallipoli in Britain from the very start.

abolished any sense of superiority or inferiority, and allowed the natural ties of common blood, similar institutions, and partial identity of interest to exert their full force. The result has been the growth of what is in a sense *a new nation or super-nation* [my emphasis], for it is composed of many constituent nations, and a new State—the commonwealth which is officially entitled the British Empire.[10]

When a similar march was staged on the first Anzac Day in peacetime, *The Times* proudly declared 'if one had to single out the distinctively British achievement among the Imperial nations of history, it would be this power of giving birth to new nations, bound to us by the closest of bonds yet with marked individuality of character, enriching the parent stock by its differences as well as by its similarities'.[11]

On the Sunday following Anzac Day, Bury, home of the Lancashire Fusiliers who won six VCs in the landings at W Beach, held its own commemoration. The rector of Bury, John Hill, noted the recent Westminster Abbey service commemorating the Anzacs in his sermon, then continued:

> But, men of Lancashire, while we pay the honour due to those gallant men who came from the dominions beyond the seas to fight and fall on our behalf, we don't and dare not forget the part played on the same day and on that same shore by the sons of England, Scotland, Wales and Ireland and, as in duty bound, we especially commemorate the immortal heroism shown by the officers and men of the 1st Battalion, Lancashire Fusiliers, our own regiment, whose home is here in Bury, whose colours hang proudly in this fair church, whose memorials speak to us from these walls. If they did not find mention in the Abbey, at least they shall find it here, for they are ours, our brothers and our flesh, and the record of their deathless deeds is the heritage of our country and of our town for all time.[12]

So began the Bury tradition of 'Gallipoli Sunday'. There was also a service and parade six weeks later in Manchester.[13] This 'Gallipoli Day' was to commemorate the anniversary of the Third Battle of Krithia (4–6 June) in which Territorial soldiers from the Manchester Brigade of the 42nd (East Lancashire) Division played a prominent part alongside Lancashire Fusiliers from Salford, Rochdale, and Bury. By one estimate, at least 2,400 men from Manchester (broadly defined) had become casualties at Gallipoli. On Monday 5 June 1916, there was a military parade at Belle Vue Gardens, followed by lunch for the men courtesy of the Lord Mayor, and then the soldiers proceeded to Albert Square, where a march past was held. The men were cheered en route, but the scene was quieter in the square where the men were greeted by the waving of handkerchiefs and a military band. Thereafter, the soldiers, who were all convalescing Gallipoli veterans, and 1,000 of their relatives attended a service in Manchester Cathedral, alongside the Lord Mayor of Manchester and the mayors of other Lancashire towns.[14] The following month, the towns of the Scottish Borders marked 12 July as the anniversary of the Territorials of the 4th and 5th King's Own Scottish Borderers' engagement at Kereves Dere.[15]

Manchester's local habit of marking 'Gallipoli Day' in June did not significantly influence the terminology applied to 25 April, despite occasional suggestions to the contrary. In 1917, an 'Anzac Correspondent' argued that England (Ireland and elsewhere were overlooked) should call the anniversary of the landings Gallipoli Day: calling it Anzac Day here would be a 'mistake': 'Let us, the Colonials, keep our Anzac day, but let us take part in the bigger and more historical Gallipoli Day from now onwards.'[16] Indeed, *The Times*'s reports of the commemorative activities of 25 April 1917 do describe it as 'Gallipoli Day for the English troops and Anzac Day for the Australians and the New Zealanders'.[17] It was a step towards greater accuracy and inclusivity, and appropriately so for a day that saw the unveiling of the most important British memorial to the campaign, that to the 29th Division at Holy Trinity Church in Eltham, south London. This was the inspiration of Holy Trinity's vicar, Reverend Henry Hall, who had served as the 29th Division's chaplain at Gallipoli. And yet the habit of the times that saw English and British used interchangeably in discussing these commemorative arrangements, implicitly excluded the vital part that the Irish soldiers of the 29th Division had played.

In 1918, 25 April was dubbed 'Anzac Day' in *The Times* once more, a day in which 'Not unmindful of her own dead, England joins in the tribute'.[18] Among the day's activities were 'pilgrimages' to 250 different locations across the country where Australian and New Zealand soldiers have their burial places. These tributes by expatriate Anglo-Australians were organized by the Australian Natives' Association. An expression of grief, a homage to sacrifice, and a substitute for loved ones buried overseas: these pilgrimages also reflect the overlapping layers of identity that existed. They expressed a distinctive emergent national sensibility while assuaging 'the common grief of empire'.[19] A pattern therefore emerged, and continued in the interwar press, of tending to call 25 April Anzac Day and focusing on Australian and New Zealand commemorative activities in London, with reports thereafter of activities at Eltham, and the Anzac grave pilgrimages.[20] This Anzac-centric commemoration of Gallipoli was epitomized by the initial decision to endorse the Dominion governments' proposal to award a medal to soldiers from Australia and New Zealand who had served at Gallipoli, without extending the honour to British, Irish, or Indian participants in the campaign. After official protests from Manchester City Council and two other Lancashire town councils, the decision was corrected.[21] No Gallipoli medal was awarded and instead all participants were eligible for a 1914–15 star.

Outside London, local commemorations linked to the local regiment or to local personnel continued in Bury, Manchester, the Borders, and elsewhere.[22] In Bath, the Lord Mayor organized a parade and reception for any veterans of the campaign on 'Gallipoli Day' until 1919.[23] By 1924, it had been replaced by a brief service at Arnos Vale Cemetery where several Anzac soldiers are buried.[24] In 1921 and 1922, while the Munster Fusiliers were based in Plymouth, their role in the landings at Gallipoli was commemorated in the cathedral, and in 1928 the Hampshire Regiment paraded at Crownhill Barracks in the same city.[25] Whereas in Australia, communities across the country marked the nation's sacrifices on 25 April, the limited commemoration of Gallipoli in Britain only reflected local connections to the campaign, and as often as not those were Anzac connections.

The Aftermath of Defeat

In hindsight, Gallipoli was the most egregious defeat for the British Empire of the entire war. Such was its magnitude that it contributed to the downfall of the last ever Liberal government of the United Kingdom. It finished the active career of its commander-in-chief, General Sir Ian Hamilton, and it eclipsed the career of its main author, First Lord of the Admiralty Winston Churchill. Some 390,000 Allied casualties[26] were suffered for no immediately apparent gain. With such significant ramifications, was it acknowledged as a defeat? The answer is yes—but in Germany, not in Britain. The Times and a handful of provincial newspapers reported a comment by the Reichstag president following the first stage of the evacuation. He described Gallipoli as 'a military and moral defeat for England'.[27] However, in Britain, it appears that it was not until at least 1928 that Gallipoli was publicly acknowledged as a defeat—and then only in passing in some book reviews.[28] Instead, the habit emerged of acknowledging a 'failure' but insisting that it was glorious nonetheless. We can trace this habit in public discussions of the campaign: in official reports, in commemorative speeches and memorials, in newspaper editorials and book reviews, and in books on the subject.

Strategic judgements on the campaign—was it a good idea? what did it achieve? what went wrong?—are relevant to all nations that fought at Gallipoli. But they were debated most fiercely in Britain because the imperial government in London was ultimately responsible for strategic decision-making. They are vital because they inform the answer to the heart-wrenching question that is surely asked of all who fell, 'Did he die in vain?' Yet these are difficult questions, and firm answers to them were rare in the first five years after the campaign. More frequent were ad hominem attacks in the search for scapegoats. The ongoing wider war inhibited the vehemence of some criticisms. While there was fighting on the Western Front, public faith in the conduct of the war had to be maintained. The nature of the fighting on the Western Front is also important here—cold, muddy, and murderous on an industrial scale—the Gallipoli campaign provided a stark contrast, where there was still scope for heroic dash under a blue sky despite the significant casualties and the horrific sanitary conditions. This meant it was possible and preferable to draw upon a pre-existing vocabulary concerning the glories of war.[29] Hence there emerged a heroic-romantic myth of Gallipoli that set it apart from the rest of the First World War.[30] It can be traced through the writings of John Masefield and those of many other British authors from the earliest days of the campaign.[31]

This romantic understanding of Gallipoli finds its origins in the British education system, particularly in its elite public schools, which emphasized the study of ancient Greece and stressed the value of chivalry, sportsmanship, and athleticism. Hence British participants in the campaign were quick to link it to the classical associations of Gallipoli and its surrounding area: Leander swimming the Hellespont to visit Hero (repeated by Lord Byron in 1810); the Persian king Xerxes and his bridge of boats; the proximity of Troy. This mindset was exemplified by, but not limited to, a group of gilded youths who served in the Royal Naval Division (RND) at Gallipoli. Among them were the prime minister's son, 'Oc' Asquith, and Patrick Shaw-Stewart who, like a number of his contemporaries, was inspired to write poetry which drew parallels with their Homeric forebears. In June 1915, for example, he wrote,

> Was it hard, Achilles,
> So very hard to die?
> Thou knewest and I know not—
> So much the happier I.

> I will go back this morning
> From Imbros over the sea;
> Stand in the trench, Achilles,
> Flame-capped, and shout for me.[32]

The most famous of this RND group was Rupert Brooke. His poem epitomizing the romance of war, 'The Soldier', had been read by Dean Inge at St Paul's Cathedral on Easter Sunday.[33] Before he could experience battle for himself, he died on St George's Day, 23 April, from septicaemia arising from an insect bite. His friends buried him, by torchlight, on Skyros, the island closely associated with the legend of Achilles. Another RND author, A. P. Herbert, wrote *The Secret Battle* in 1917. This searing indictment of the court martial procedure uses the romantic setting of the Gallipoli campaign as a means to establish the idealism and bravery of Harry Penrose, a young officer whose nerves were subsequently worn down on the Western Front.[34] With an introduction by Winston Churchill, the book was in its ninth edition by 1949.

Thus, the immediate response to the campaign was a romantic reflex in lieu of a considered strategic judgement. After the well-executed but essentially humiliating evacuation, the *Devon and Exeter Gazette* chose the headline: 'Gallipoli. Heroes' Imperishable Fame. A Miraculous Evacuation. "Without Parallel in Military History."'[35] London's *Daily News* considered that 'The brave men who have given their lives by thousands on these barren shores have not died in vain. Their fame is already secure.'[36] A more sorrowful *Manchester Guardian* focused on the fact that in this second evacuation 'The Turks [had been] Again Outwitted',[37] and indeed much of the positive reporting resided in admiration for the exceptional feat of withdrawing without loss of life.

At commemorative events, the exaltation of heroic sacrifice was far more appropriate than an inquest into politics and strategy—the latter simply would not do for an occasion which had to honour the fallen, assuage grief, and, in wartime, maintain morale. Thus, for example, the use of rhetoric rooted in the perceived relationship between warfare, masculinity, and nationhood. *The Times's* editorial on the first Anzac Day commemorations said,

> Never, to all appearance, was valiant blood spilt for less gain; never 'the shield of the mighty' more 'vilely cast away.' Yet we know that it was not so. The Australians and New Zealanders, when they landed in Gallipoli, set the seal upon the manhood of their peoples.[38]

During the service at Westminster Abbey, the dean of Westminster declared that the men had 'died at Gallipoli for their King and Empire, in the high cause of Freedom and Honour'.[39] These capitalized abstract nouns are appropriate for the soothing balm of a commemorative occasion. The essential and consistent part of all such moments was that the heroism of the men who fought was asserted and celebrated. Writing of the Manchester Territorials' actions on 4 June 1915, the *Manchester Guardian* wrote, 'it is to-day, a year afterwards, and in like years to come, that the real harvest is reaped—the glory, the undying example, and the courage which will never submit or yield even in the face of certain death or hopeless enterprise'.[40] In 1917, *The Times* added a rebuke to the criticism and political controversy that had begun to swirl around the campaign: 'No Royal Commissions, no censure of political parties, no lying *communiqués* from the enemy can ever rob the enterprise of its wonder.'[41]

This was a reference to the Dardanelles Commission appointed to inquire into the campaign. The publication of its first report in March 1917 was a moment for some limited assessments. At this point *The Times* acknowledged it had been a 'gigantic failure'.[42] Yet, like most other papers, it did not express an opinion on the strategy. Only upon the post-war publication of the final Dardanelles Commission report did it state, 'it is now generally agreed that the strategic idea of the Dardanelles was sound'.[43] By contrast, the *Western Times* considered in 1917 that 'the whole story constitutes a tragedy of supreme importance, for the opening of the Dardanelles would probably have by this time ended the war'.[44]

In place of considerations of the overall strategy, the newspapers followed the report in criticizing process within the War Council—and there were various abject failings to criticize—and then developed their ad hominem attacks on a range of politicians and officials, all of whom were now safely out of office. Aberdeen's *Evening Express* criticized the secretive and overburdened Kitchener, but reserved its greatest censure for Fisher.[45] Dundee's *Evening Telegraph* noted that Asquith, Kitchener, and Fisher had all been criticized for 'the Dardanelles Muddle'.[46] For *The Times*,

> Mr Churchill remains, as the public have rightly held, the prime mover in the Dardanelles adventure. He was at least consistent in his purpose when all the rest were vacillating. But it was the consistency of a dangerous enthusiast, who sought expert advice only where he could be sure of moulding it to his own opinion, and unconsciously deceived both himself and his colleagues about the real character of his technical support.[47]

By contrast, the *Manchester Guardian* drew the conclusion that the report 'removes the greater part of that unjust burden from his shoulders. [...] [he was] the driving spirit in the whole enterprise and that is to his credit. But it is not true that Mr. Churchill overrode his experts.'[48] Later that year, as Churchill faced a by-election in Dundee prompted by his return from the political wilderness to be minister of munitions, the newspaper was prompted to defend him again from 'the bitterness with which a small section of his critics pursue him, in Parliament and in the press'.[49]

This stance was in keeping with the unusual editorial line of the *Manchester Guardian* in consistently expressing very clear support for the strategy, and making large claims for its impact on the war overall. On 11 April 1916, an editorial wrote of the 'failure of the really fine strategic idea of striking at the very heart of Turkey. [...] The campaign promised to shorten the war by a year; the failure has probably lengthened it by a year.'[50] It also argued that 'the war might have been won' at Gallipoli.[51] By 1921, it was suggesting additionally (and implausibly) that failure at Gallipoli had 'caused the downfall of Russia', which is to say that it caused the Russian Revolution.[52]

Britain and the Memory of Gallipoli in the 1920s and 1930s

The patterns established in wartime continued over the following two decades: a romantic view of the campaign, the commemoration of local British connections, and a general tendency to view Gallipoli as Anzac business. Indeed, the early 1920s should be seen as the high-water mark of the British heroic-romantic myth that defended the campaign, beginning with the warm reception afforded General Sir Ian Hamilton's *Gallipoli Diary* in 1920.[53] His despatches had already been published in newspapers, as a book, and as an appendix to the Dardanelles Commission's final report.[54] Still feeling that his name had not quite been cleared, he published his memoirs in diary form. For a military man, Hamilton had a somewhat grandiloquent style of writing, and his account was shot through with references to the campaign's classical setting and his own deeply romantic view of

warfare which stressed its chivalrous and gallant aspects, while sup-
pressing its horrors. Meanwhile, the book systematically asserted the
strategic potential the campaign had to shorten the war by up to two
years. It also sought to show how Kitchener and the politicians had
failed to properly support him with armaments and reinforcements.
Meanwhile, Hamilton assiduously unveiled memorials and attended
regimental dinners, making speeches celebrating the campaign on
many occasions.

The most excessively romantic portrayal of Gallipoli, Ernest
Raymond's *Tell England* (1922), was probably also the most popular:
it sold 300,000 copies by the end of 1939, and by 1965 was in its
fortieth edition.[55] The author had been the chaplain to the East
Lancashire Territorials of 42nd Division, and his story traced the
experience of two idealistic schoolboys en route for what they saw
as a crusade to rescue Constantinople for Christendom. The book's
theme is the adventure of war and the glorious spiritual beauty of
dying in the name of patriotism. Its title was drawn from the epitaph
on one of the character's grave, and was inspired by Simonides'
epitaph on the Spartans who fell at Thermopylae:

> Tell England, ye who pass this monument,
> We died for her, and here we rest content.[56]

This was a reassuring and lively version of war as it should have been.
Interestingly, its cinematic version (released in 1931), which caused the
IRA such offence, reflected its times with a less romantic and more
disenchanted view of warfare.

The year after the novel was published, Churchill's memoirs were
serialized in eighteen articles by *The Times* in October 1923.[57] *The World
Crisis 1915* was his opportunity to defend himself at length from the
sustained criticism he had faced since the campaign's inception. He
had made his case vigorously before the Dardanelles Commission, in
speeches and in newspaper articles. Now he drew on his literary gifts
to defend his eastern strategy and to put the boot into Kitchener for

hobbling the campaign. He included extracts from many contemporary documents—sometimes misleadingly edited—and artfully glossed over difficult issues, invoking Fate to explain away failure.[58] He was highly persuasive in his claims for the potential of his strategy. Among the sympathetic reviews he garnered, *The Observer's* made the greatest claims:

> It would have shortened the world war by two years, and saved millions of lives, while preventing the submarine threat from growing in 1917 to the deadly menace it became. Halving the burthens [sic] of taxation and unemployment that press upon us now, it would have prevented the collapse of Russia, the total shattering of organised Europe, and that peace of chaos which has followed the war of exhaustion.[59]

Churchill also had a hand in enabling the British official historian of the campaign to express criticisms of the government and its failure to prioritize Gallipoli ahead of the Western Front. Cecil Aspinall-Oglander, a staff officer at the campaign, brought to his task as official historian a sense that the campaign's value needed to be explained and its failures justified. Thus despite the constraints of a genre that in Britain was primarily concerned with military education, his volumes (published in 1929 and 1932) tended to emphasize leaders rather than battle plans as an explanation of failure. He scapegoated Stopford, made some limited criticisms of Hamilton, and highlighted the calibre of his opposition by praising Mustafa Kemal. Aspinall even laced what should be a dry study of strategy and tactics with some limited references to the romance of the campaign. The heroic-romantic myth pervades this part of the British literature of the campaign too.[60]

Yet romantic perceptions of the campaign did not translate into widespread active engagement in its commemoration. For the tenth anniversary in 1925, a large crowd gathered in London for the unveiling of a memorial to the RND on 25 April, and the 29th Division held a reunion dinner at the Café Royal. There was a service for the Anzacs at St Clement Danes,[61] and pilgrimages to Anzac graves across the country.[62] Gallipoli Sunday continued in Bury, and Reverend Hill once more asserted Lancashire's equality of sacrifice with the Anzacs despite official arrangements:

> nor did we grudge the fact that the 25th of April was called after them Anzac Day. It was right that what they had done should be kept in mind, telling to all generations that the ties of Empire when this greatest war of all time was being waged meant much indeed, and produced something else in our hearts. Anzac Day for the Empire was Lancashire Fusiliers Day for us: our men no less than those from the southern lands achieved the impossible, and Lancashire Landing stood as their memorial to all time.[63]

In Scotland, the July commemoration of the KOSBs at Gallipoli was viewed in terms of the familiar historical yardstick of Britain's martial tradition, with Scottish and British used as interchangeable identities. As an editorial in Selkirk's *Southern Reporter* stated,

> the men of our day and generation were found to be made of as sound material as ever were Britons in the past. July 12 added lustre to British arms, and if it has meant sorrow, it has brought imperishable glory to thousands of British homes. [...] They were as true to the honour of Scotsmen as ever our men were in the old days.[64]

Beyond examples of this type, there were no major public gatherings in Britain to commemorate Gallipoli in 1925, nor for that matter in 1935 either. All commemorative events for the First World War had been rolled into Armistice Day each November. Instead these limited public commemorations of Gallipoli were augmented by private commemorations such as old soldiers' reunions. One such occasion was the gathering in October 1931 of former members of 'E' Company, 4th Battalion, Royal Sussex Regiment to view *Tell England* at the cinema in Horsham, followed by a reunion dinner featuring dishes named after places in Gallipoli.[65] The campaign was important to those with direct personal connections to the campaign, or for its Anzac aspects. Gallipoli did not have its own national resonance in Britain.

Ireland and the Memory of Gallipoli
in the 1920s and 1930s

The tendency to romanticize the campaign was not unique to Great Britain. Given the shared cultural heritage of the British Empire, it is unsurprising to find that before they developed their distinctive Anzac voices, Australian and New Zealand newspapers were quite as ready as their British counterparts to draw comparisons with ancient defeats at Thermopylae or to quote extensively from the mellifluous prose of John Masefield.[66] Although Irish newspapers do not appear to have reached for classical allusions in the same way, the heroism of the Irish was celebrated as romantically as anywhere else. They shared with Australian reports a tendency to read into events innate martial qualities resulting from their nationality.[67] Indeed, save for the description of the men's religiosity, the *Irish Times*'s review of Bryan Cooper's book *The Tenth (Irish) Division in Gallipoli* could fit happily into the Anzac legend—the men dash forward in battle, bayonets ready, they are tenacious in defence, and there are flashes of humour:

> Appeals to Allah inspirited the Turks, but did not impress the Connaught Rangers. 'They came on shouting and calling for a man named Allen', said one, 'and there was no man of that name in the trench at all.'[68]

The Irish public were proud of their soldier heroes: when the Protestant chaplain in the 10th (Irish) Division declared at a Limerick County Council meeting in March 1916, 'He had never felt prouder of being an Irishman than he was when he saw those gallant fellows fighting with dash and heroism against terrific odds',[69] he was greeted by applause. When Sergeant James Somers of Cloughjordan, Co. Tipperary was awarded the Victoria Cross in 1918 for his bravery at Gallipoli, a collection in the county raised almost £300 to present to him in celebration.[70]

Nonetheless, there were conflicting currents within Irish reactions, suggested, for example, by Francis Ledwidge's poem, 'The Irish in Gallipoli', written in February 1917. Ledwidge, a nationalist who none-theless served in the 10th Division at Suvla, grapples with justifying his service in the British Army amidst 'The Threatening splendour of that isley sea | Lighted by Troy's last shadow'.[71] Elsewhere, romantic sentiments in Irish newspapers were accompanied by harsher judge-ments on the campaign and in particular on politicians in London. The day after the final evacuation of Cape Helles, the *Irish Independent* described Gallipoli in its headline as 'Tragic Chapter's End. Year's Useless Campaign'.[72] The Unionist *Irish Times*, in its editorial on General Monro's despatch, noted, 'We have abandoned Gallipoli, but it remains populous with memories of a reckless statesmanship and of heroic deeds.'[73] Sir Thomas Esmonde MP, chief whip of the Nationalist Party, went much further, arguing that the 10th (Irish) Division had been 'recklessly thrown away by British incompe-tency'.[74] Later that year after reviewing Masefield's defence of the campaign, the *Irish Times* nonetheless concluded, 'If blame be due, we are not yet in a position to apportion it. It is certainly not our place to criticise the leaders of the great adventure, or to find in it nothing but a ghastly failure.'[75] Such was the currency of this view of the cam-paign that the following year an ardent nationalist used Gallipoli as a metaphor for disaster.[76]

There was also a growing sense that the role of the Irish had not been properly acknowledged in the first place. During the war, the *Irish Times* praised the publication of a book entitled *The Irish at the Front* as the answer to the secrecy surrounding the activities of the Irish regiments.[77] One angry correspondent felt that the 10th (Irish) Div-ision at Suvla had been not only forgotten but also blamed for the blunders that occurred there;[78] another suggested that the division was badly treated throughout the war.[79] One of the difficulties in giving the 10th Division their due was that they never fought all together at Gallipoli, and so it was difficult to tell their story or ascribe one particular action to them.[80] There was also no one to tell it: no

Irish war correspondent accompanied the MEF,[81] nor did official communiqués make up the deficit. The *Irish Independent* complained that General Sir Ian Hamilton's 'despatches gave very scanty recognition to the valour and sacrifice of the Irish battalions in his command'.[82] Indeed, not even the outstanding role of the Irish soldiers in the landings at V beach was mentioned in official despatches.[83]

I have not found any Irish commemorations of the campaign during the war. The first anniversary of the regular Irish soldiers' role on 25 April was not commemorated in 1916. The Dublin-based *Irish Times* only reported the arrangements at Westminster Abbey for the Anzacs,[84] while the *Impartial Reporter*, local paper for the recruiting area of the Inniskilling Fusiliers in the north, devoted its 27 April editorial to condemning the ongoing Easter Rising rather than noting the anniversary. Similarly, no August tradition of commemorating the 10th Division at Suvla seems to have emerged. Yet the Anzac connections *were* noted, such that by 1919 the main Anzac Day activity was the 'usual' pilgrimage to Anzac graves in Dublin.[85]

This is not to say that there was no commemoration of the role Irishmen played in the First World War—they were simply too numerous to be ignored or suppressed—but the political atmosphere was not necessarily encouraging. At the end of the war, elected Irish representatives declared their nation's independence, to which Britain responded with brutal violence. The ensuing War of Independence, which resulted in the partition of Ireland in May 1921 and the establishment of the Irish Free State in December 1922, was followed by civil war, which concluded in May 1923. In this context, the 1919 appeal for funds for an Irish National War Memorial, despite stressing the Irishness of those who had fought in the First World War, struggled to raise sufficient funds or find a suitable location. It was eventually located on the outskirts of Dublin and partially completed in 1938 but not officially opened before renewed war intervened. Although the Irish government was not hostile to such commemorations, it did not care to promote them. Furthermore, two of the main institutions that might have officially sustained the memory of Gallipoli disappeared with the disbandment of the Dublin and Munster Fusiliers in 1922. Thus following 'Anzac Day' in 1922, the Anglophile *Irish Times* asked sadly, 'Did anybody lay an Irish wreath yesterday on the Cenotaph in London?'[86] The November commemorations were very well attended, but fraught with controversy at the periphery. Poppy-snatchers would grab this perceived symbol of Britishness from its wearers, and carry away and burn any British flag they could.

Those with a personal connection did commemorate Gallipoli. 'British' ex-servicemen, largely of the now disbanded Munster Fusiliers, staged a march in Cork each year.[87] In 1929 they numbered around 300.[88] Such commemorations were important to those with a personal connection, but no wider lesson was drawn by politicians or the public at large; quite the contrary: it was to be forgotten. The *Irish Times* bemoaned the fact that the example of the Dublins and Munsters, who 'were the supreme heroes of the deathless day', could not be used to inspire national pride in Ireland as it did in Australia. In an editorial entitled 'Pro Patria', it wrote:

> In a message to Australia yesterday Mr. Bruce, the Commonwealth Prime Minister, said that the deeds of the 'Anzacs' at Gallipoli had inspired all their fellow-countrymen with 'a wider conception of nationhood.' Who can believe that the Dublins and Munsters died in vain, even though Irish statesmen of smaller vision than Mr. Bruce still dispute their claim to their country's reverence?[89]

By the twentieth anniversary in 1935, the *Irish Times* was espousing something like an Irish heroic-romantic myth, bearing all the hallmarks of the British myth—classical references, the celebration of heroism, an assertion of Gallipoli as a war-winning strategy—but additionally featuring a sideswipe at 'bungling politicians' and an extended claim for the Irishness of the campaign: Admiral De Robeck was an Irishman, Sir Roger Keyes 'the heart and soul of the attack, is a son of Ulster', and above all the men themselves. It noted the 'imperishable fame' of the Dublin and Munster Fusiliers and the 'proud achievements' of the 'Pals' Battalion and other Irish troops on the peninsula, plus the fact that very many Anzacs were Irish emigrés. Nonetheless, they had been ignored:

> Those troops knew no difference of religion or politics. Unionists and Nationalists, Protestants and Catholics—they all were proud to wear the King's uniform, and to die in it; yet to-day, while London gives them unstinted praise, their own country does not seem to think them worthy of a thought.[90]

We see here the bones of what might have been, an Irish legend of Gallipoli, but the romantic view of the *Irish Times* did not resonate more widely in Irish society. Indeed, that this was a minority view can be inferred from the fact that no extended references to First World War commemorations have been found in Irish newspapers other than the *Irish Times*.

In the 1930s, what limited support the Irish government had lent to First World War commemorations was further reduced. With the election of the first Fianna Fáil government in 1932 (which took a more militantly republican point of view), representatives of the government no longer attended commemorations within the country, Union Jacks and militaristic parades were banned from commemorations, and the sale of poppies was restricted. As historian Philip Orr explains: 'Thus did Eamon de Valera's government seek to allow commemorative space for those who publicly wished to mourn the dead of the Somme and Gallipoli whilst striving to rid the remembrance process of its pro-British aura.'[91]

· Thus in Ireland, although the war was commemorated each year, it could be controversial and it was certainly not part of a nation-building process as in Australia. Service in the British Army was not an appropriate vehicle for this. As the *Irish Times* recognized in connection with the difficulties of the Irish National War Memorial scheme, 'The [First World] War is no part of the chapter of Irish history which began in 1916; and the Irish dead of Flanders and Gallipoli have no imperious claim upon the gratitude of the new regime.'[92] Meanwhile, commemoration in Northern Ireland never focused on the Inniskillings on X beach at Gallipoli, nor the northern men who'd joined the 10th Division. Instead the province looked to France and Flanders, and particularly the mainly Protestant 36th (Ulster) Division's role at the Somme as a means to stress its Britishness.[93]

The Memory of Gallipoli in Britain and Ireland after the Second World War

During the Second World War, Ireland was neutral. To avoid the suggestion of support for the British war effort, parades on Armistice Day—given their ritualistic British trappings—were banned in Dublin.[94] Thereafter, commemoration of the earlier war largely fell into abeyance in Ireland—symbolized by the dereliction of the Irish National War Memorial. Dublin's Grangegorman cemetery, which had once been visited by pilgrimages on Anzac day, was reported as being 'in very poor order' in 1958, the weeds only kept at bay by an elderly caretaker with a scythe. In the 1970s, some war graves in Ireland were vandalized, and by 1981, 116 Commonwealth War Graves Commission sites were overgrown with brambles.[95] It was unacceptable to acknowledge Ireland's participation within the British Empire's forces, particularly so amidst the violence over Northern Ireland's sovereignty between 1968 and 1998. Yet it was possible to acknowledge Australia and New Zealand's role, and observance of Anzac Day continued throughout much of the post-war years. The Australian connection in the congregation and the celebrant was always noted, perhaps as a means to deflect any seeming British connection.

Thus in 1947 the Anzac Day Mass at St Kevin's Oratory, Pro-Cathedral in Dublin was conducted by Reverend Charles Haughey, O.Carm., who had been a chaplain in the Australian forces.[96] When Reverend P. J. O'Farrell OSA took over in 1955, it was noted that he had 'spent a number of years in the Diocese of Cairns in Northern Queensland'.[97] Then an Australian priest, Reverend Kevin Walsh, officiated in 1956.[98] By the mid-1960s, the service was held in the Australian

147

Embassy itself. There is a gap in the records after 1972. Possibly no services were held in the following fifteen years, then occasional events resumed (or their notification in the newspapers resumed). In 1987 there is a small note of an Irish Australian Society Memorial Mass on Anzac Day at the University Church, St Stephen's Green, Dublin.[99] In 1993 the New Zealand Ireland Association attended choral evensong in Dublin for Anzac Day. In 1995, the Irish–Australian Society (which had been wound up in 1977) marked the eightieth anniversary with a memorial mass. From 2006, the Australian Embassy began to organize a dawn service in Dublin.

For different reasons, a similar situation applied in Great Britain. In 1965, an Australian newspaper lamented that 25 April was not marked in Britain: 'Anzac Day has no special significance for the English—not even as Gallipoli Day.' In London, the wreaths on the Cenotaph and the service at Westminster Abbey were organized by Australia House: 'Left to themselves, the British would hardly bother.' But this was not specific to Gallipoli or other individual campaigns of the First and Second World War. As the Australian journalist noted, not even Dunkirk was given special treatment: 'The coldly logical official attitude, explained to me this week, is: "After all, old man, if you do one, you have to do the lot, and it just isn't on."'[100]

This is not to say that the British were entirely uninterested in Gallipoli at this stage. Publications on the First World War across Europe and America were at an all-time low in the 1940s and 1950s,[101] but Alan Moorehead's *Gallipoli* (1956) prefigured the growing interest in the subject from the 1960s. Writing with a journalist's eye for an interesting story and a vivid turn of phrase, Moorehead defended the campaign and its strategic potential. Indeed, with Winston Churchill's reputation transformed by his leadership in the Second World War, and recent examples of the difficulties and heroism required in amphibious landings (D-Day) and evacuations (Dunkirk), Gallipoli was seen in a new light. Moorehead's *Gallipoli* sparked off romantic flights of hyperbole in reviewers of the book, won two major awards, and was the basis for a programme scheduled for ITV's first night of broadcasting.[102] It also inspired Robert Rhodes James to write his classic book on the campaign, published in 1965.

The response to Rhodes James shows us something about the different attitudes informing the seeming forgetfulness in the British Isles. Book reviews in Ireland were far more vicious than their British counterparts. That in *The Times* was brief and bothered only by the campaign's strategic implications. It concludes with a certain indifference: 'Mr. James has shaped a mass of material into a coherent, dramatic and controversial narrative. He will open up old wounds.'[103] The unrepentant controversialist A. J. P. Taylor summarized the boldest criticisms of the book in *The Observer*: 'Gallipoli was an example of the wrong troops, fighting under the wrong generals on wrong plans in the wrong place.'[104] And yet, even he appeared a little wistful in his conclusion: 'Still, when one counts up the lost opportunities, it is difficult to resist a feeling that the enterprise should have come off at one point or another.' Contrast that with the unflinching focus on the cost of the campaign (and the partial acknowledgement of some soldiers' heroism) in an Irish review of the same book:

> Quietly, it terrifies. Here is the black heart of all that heroism by Australian and Gurkha, Dublin and Munster Fusiliers; here behind the war that poets and novelists tried to romanticise with visions of ancient nearby Troy of the heroes, here is the reckoning that for eight-and-a-half months littered beaches and hillsides with dead. The sea turned crimson. Dysentery and other diseases went on the rampage. Men ran into machine-gun fire. The wounded rotted under shrouds of flies in the sun. Turks and Allied soldiers knew profound pity for one another.[105]

To the limited extent that the British remembered the campaign, its romantic aura remained. For the Irish, it was angrily forgotten.

Modern-day Gallipoli in Britain and Ireland

In recent decades, the context in which Gallipoli is remembered has changed—in Ireland's case, remarkably so. As the last veterans of the

First World War dwindled, a sense of urgency emerged in attempting to capture their experiences. Popular history books and novels about the war became bestsellers at the turn of the millennium and, along with the growth in family history, became 'part of a burgeoning sub-culture of fascination with the war'.[106] With the possibilities for discussion and access to original documents offered by the Internet, and the approaching centenary with its attendant media coverage, interest in the First World War is probably higher in Britain than at any time in the last fifty years. Interest in Gallipoli, as one of its most significant campaigns, has also grown. Membership of the Gallipoli Association has almost trebled. Once a veterans' organization, now an association for those with a personal interest in the campaign (often sparked by a family connection), its membership has grown from 334 in 1991 (including 62 veterans), to around 900 on the eve of the centenary.[107]

Fig, 22. Members of the Fusiliers Association march to mark Gallipoli Sunday in Bury, Lancashire in 2014. This local tradition dates from 1916 and commemorates the role of the Lancashire Fusiliers who won 6 VCs when they landed at W beach at Cape Helles on 25 April 1915. The event is attended by local dignitaries and crowds of a few hundred people.

In Ireland (and in Australia and New Zealand) the same social changes apply but to these must be added remarkable political change. During the 1990s, the involvement of Irish soldiers in the war began to be acknowledged through family history, and most notably through the re-established Royal Dublin Fusiliers Association from 1997 which stressed the human experience of the war, not its politics.[108] The public reaction to the depths of the divisive politics of the conflict in Northern Ireland, including the 'Poppy Day bombing' on 8 November 1987 in Enniskillen, prompted increased attendance at commemorations across Ireland, and made apparent renewed interest in the war. Thereafter, political changes have served to detoxify Ireland's relationship with Britain and to encourage the country's remembrance of the war. These changes include: the steps towards conflict resolution between Britain and Ireland in Northern Ireland which concluded in the Good Friday Agreement of 1998, the confidence and optimism of Ireland's 'Celtic Tiger' period of economic success in the late 1990s, and perhaps, a desire to assert Ireland's role in a key moment of European history after the establishment of the single European market in 1992.[109] On 26 April 2001, at a state reception for the Royal Dublin Fusiliers, Taoiseach (prime minister of Ireland) Bertie Ahern announced, 'I am glad to say that in recent years, [...] we have had the national self-confidence to recognise that our past has many strands, and that there is not just one officially sanctioned historical canon.'[110]

The changing Irish attitude to the First World War has been signified and led by the involvement of successive heads of state and leading politicians in its commemoration. The process began in 1993 when President Mary Robinson attended a Remembrance Day service in Dublin. Among other commemorative acts over time, in 2010 President Mary McAleese travelled to Gallipoli to unveil a memorial at Green Hill Cemetery, Suvla. This was the first official recognition by the Republic of Ireland of the Irish soldiers who fought in the British Army at Gallipoli. The process culminated in Queen Elizabeth II's state visit to Ireland in 2011, which included accompanying President McAleese both to the Garden of Remembrance which commemorates those who died in the cause of Irish Independence and to the Irish National War Memorial which commemorates those who died in the First World War.

This means that Ireland's role in the First World War can now be discussed openly and without significant controversy. The country has embarked on a decade of centenaries to remember the turbulent years from 1912. But this acknowledgement and remembrance are of a quite different order of magnitude compared to Australia or New Zealand's Anzac revival. Similarly, in Britain, Gallipoli has remained a matter of personal, family, and regimental connections. It is acknowledged as an important campaign in the war, but it has no wider significance of its own. When Darcy Jones, the last British veteran resident in Britain, died on 11 January 2000, there was an obituary in the *Daily Telegraph* and a note of his death some months later in *The Times*'s Anzac Day editorial, which noted how the wreath-laying in St Paul's Cathedral 'was imbued with none of the spirit of national awakening that permeated the ceremony at Anzac Cove where the Prime Ministers of Australia and New Zealand made moving addresses'.[111] (The British government only sent a junior minister to Anzac Cove on that occasion: John Spellar, minister for the armed forces.)[112] When Percy Goring, the last British veteran (resident in Australia) died on 27 July 2001, he was buried with full military honours.[113] I am unaware of the identity of the last Irish veteran of the campaign.

Thus, in Britain and Ireland Gallipoli continues to be commemorated in the main as an Anzac occasion. In central London it has been marked since 1996 by a dawn service, originally in Battersea Park, and now at the Australian War Memorial in Hyde Park, London. A Gallipoli memorial organized by the Gallipoli Association was unveiled in the crypt of St Paul's Cathedral in 1995. Each year there is a short service there, prior to the wreath-laying at the Cenotaph, and an Anzac Day service at Westminster Abbey, both of which are organized by the Australian and New Zealand high commissions. There remain perhaps fifteen small non-Anzac commemorations of Gallipoli in Britain and Ireland outside central London.[114] There have also been one-off occasions, such as the memorials unveiled on 22 and 23 May 2005 to the men of the 1/5th Royal Scots who were en route to Gallipoli when they were killed in the Quintinshill rail disaster.[115] In 2003, there was a memorial service to John Simpson Kirkpatrick in South Shields, where there is a statue and a pub named after the man with the donkey.[116] This local man has become the most famous Anzac, and the interest in him reflects the more numerous (twenty-seven) and well-attended commemorations relating to Anzac connections within Britain and Ireland.[117] Further indication, if it were needed then, that Anzac Day is Australia's greatest export.

8

Turkey and 18 March

We are going to introduce the year of 2015 to the whole world. We
will do so not as the anniversary of a genocide as some people have
claimed and slandered, but as the anniversary of the glorious resistance of a nation, the anniversary of the resistance at Çanakkale.[1]

Dr Ahmet Davutoğlu, Turkish foreign minister, 25 April 2011

The years 2008–23 mark the centenary of a turbulent period in
Ottoman/Turkish history: the proclamation of constitutional
rule, the rise to power of the Committee of Union and Progress
(CUP), a series of defeats, and finally the War of Independence and
the 1923 establishment of the Republic of Turkey, which marks the
fall of the Ottoman Empire. In the midst of these years, 2015 is
the centenary of the Ottoman victory over Britain and France at
Gallipoli—'the glorious resistance of a nation' as the Turkish foreign
minister Dr Ahmet Davutoğlu put it in 2011—but also the centenary of
what is generally agreed in the West—or 'claimed and slandered'
according to Davutoğlu—to have been the Armenian genocide.

Two of the most important events of the First World War in the
Ottoman Empire began within hours of each other: 24 April 1915 is
marked as the beginning of the forced deportations of Armenians;
then, on 25 April 1915, a peninsula in Thrace between the Sea of
Marmara and the Aegean was invaded by British and French forces.
Historiographically speaking, the two events these actions heralded—
the Armenian genocide and the Gallipoli campaign—exist in different
universes. And only in the last decade or so has the fate of the Armenians

begun to be investigated by an increasing number of Turkish scholars. While Gallipoli has always been remembered, the genocide has been a taboo in Turkey (and when it has been used, the word 'genocide' has generated huge controversy). They are both powerful illustrations of how a society relates to and imagines its history.

Tracing the memory of Gallipoli through Ottoman and then republican Turkish history, makes clear that the context in which Gallipoli is remembered here is by far the most complex and challenging for any of the belligerent nations. For Turkey, Gallipoli is a victory amidst a series of defeats. Those defeats seemed shameful to many and so were forgotten; a logic which applies tenfold to the genocide. But they were also the responsibility of the CUP regime, and therefore in building the legitimacy of the new republic—and save for the most senior CUP leaders who had fled or been interned, there was a great deal of continuity among the leaders of the Ottoman Empire/Turkey between 1908 and 1945[2]—it was extremely useful to leave behind the broader memory of the First World War, and with it the actions of the previous government.

Thus, in building its identity the Republic of Turkey looked back to two key victories: Gallipoli and the Turkish War of Independence of 1919–22, which culminated in the republic's recognition in the 1923 Treaty of Lausanne. As a foundation myth, the War of Independence is by far the more important, but the memory of Gallipoli is nonetheless interesting and the link between the two is Mustafa Kemal. As a lieutenant colonel at Gallipoli he played a decisive role at two pivotal moments; he then went on to lead the Turkish national movement which fought to overthrow the stipulations of the Treaty of Sèvres, end the Ottoman sultanate, and establish sovereign, secular, and democratic government in Turkey. He became its first president and remained in power until his death in 1938. Notably, Kemal's role at Gallipoli became significantly more acclaimed after he attained power. Prior to that, it was the humble soldier who was primarily celebrated for his heroism at Gallipoli. Known affectionately as Mehmet or Mehmetçik, he is the most consistent facet of the Ottoman/Turkish memory of the campaign. Kemal rises from relative obscurity in historical memory to dominate the story in the 1920s and 1930s, and later returns to centre stage from the 1980s onwards.

An Inspirational Victory in Wartime

For decades, European powers had been jockeying for influence in what was perceived to be the crumbling Ottoman Empire. This sprawling multi-ethnic entity was ruled by a sultan who was also the caliph—that is, leader of the world's Sunni Muslims. His government was known as the Sublime Porte. After the CUP military *coup d'état* in 1913, the sultan remained as a figurehead, but real power resided in the triumvirate of Enver, Talat, and Cemal Pashas. Humbled in wars in the Balkans before 1914, the Ottoman Empire declared in favour of Germany and Austria-Hungary in the First World War and fought on multiple fronts along and beyond its vast borders: the Caucasus, Sinai-Palestine, Mesopotamia, Hijaz, Gallipoli, Macedonia, Persia, Galicia, and Romania. It recruited 2,850,000 soldiers,[3] of whom approximately 800,000 died during the war. Disease and starvation also took a huge toll, elevating deaths among Anatolian Muslims to 2.5 million. To this must be added the impact of massacres and forced migrations on the Armenian and Greek populations. Overall, the population of Anatolia was reduced by 20 per cent through war, disease, and starvation.[4] With the exception of the two decisive victories at Gallipoli and Kut al-Amara (in Mesopotamia/southern Iraq), the Ottoman army suffered an almost uninterrupted string of military defeats during the war.

By the beginning of 1915, the Ottoman army had already incurred terrible losses at the hands of the Russians and of the weather in the freezing conditions of the Caucasus Mountains in the north-east of the empire. Then from February, it faced an attack more than 1,000 miles to the west from the combined forces of the British and French empires. This was the naval assault on the Dardanelles, the narrow straits which guarded the route to Constantinople, capital of the

Ottoman Empire. The British halted their activities on 18 March in order to await the readiness of the Mediterranean Expeditionary Force (MEF) for a joint amphibious assault. What was to them a pragmatic decision prior to a renewal of the fight came to be seen by the Ottomans as a stunning naval victory. The imperial capital had been saved from invasion, and two other empires had been humbled. Gallipoli became a source of national pride and inspiration, and was promoted as such in propaganda which celebrated the bravery of the ordinary soldier.

For the first anniversary of the Ottoman Empire's entry into the war on 29 October 1915, the newspaper *Tasvir-i-Efkâr* put a symbolic image of an infantry soldier at the centre of its front-page illustration. To his left was an image of Cevat Pasha, commander of the victory at sea on 18 March, and to his right was Mustafa Kemal. The caption read,

> Colonel Mustafa Kemal, brave and courageous by nature, who was extremely effective in the ground war at Gallipoli, and one of the commanders who, with the forces under his command and with great skill, gloriously saved the straits and the seat of the caliphate.[5]

Thus the ordinary Ottoman soldier was given the greatest prominence in defending the caliphate—a religious emphasis that was later dropped, but the importance of the naval victory and Kemal's role were also highlighted. But when the sultan officially decorated the 27th and 57th regiments that resisted the Anzac landings on 25 April, Kemal went unrewarded; nor was he mentioned in the official communique of the ceremony published with photographs in *Harb Mecmuası* (War Magazine) in April 1916.[6]

Shortly after the campaign, Sultan Mehmed V. Reşad (r. 1909–18) composed a *gazel* (lyric poem) about the campaign in which he lauded the courage and strength of 'his' soldiers who had safeguarded the 'heart of Islam', i.e. Constantinople:

> United did two of Muslims' mighty enemies
> Assail Gallipoli by sea and land.

Yet did God's aid reach our army
And our every soldier a steel-framed fort become.

At last did the enemy accept his impotence
Faced with our soldier sons' resolve.

Coming to seize the heart of Islam
He fled with his honour in the dust.

Reşad! Cover your prayer rug in thanks
And pray God ever guard the Muslim lands.[7]

Thus did the caliph celebrate the heroism of Ottoman soldiers against 'two of Muslims' mighty enemies'—that is, Britain and France—in strongly religious terms. The poem was published in a range of contemporary Ottoman newspapers in August and September 1916.

There was also a concerted effort to encourage prominent intellectuals, journalists, and artists to create artistic and literary works in response to the campaign. At the behest of Enver Pasha himself, via the General Staff Headquarters Intelligence Office, a large group departed for Gallipoli from Constantinople's Sirkeci train station on 11 July 1915. These were writers and journalists Ağaoğlu Ahmed, Ali Canip, Celal Sahir, Enis Behiç, Hakkı Süha, Hamdullah Suphi, Hıfzı Tevfik, Muhittin, Orhan Seyfi, Selahattin, Mehmed Emin, Yusuf Razi, Ömer Seyfettin, İbrahim Alaeddin, and Müfit Ratip; the musician Ahmed Yekta; and the painters İbrahim Çallı and Nazmi Ziya.[8] Their invitation stated that the aim was to make the soldiers' sacrifice 'live forever in national memory and to raise the new generation with these feelings'.[9] After the trip, many members of the group published articles and poems about their Gallipoli front experience in famous contemporary newspapers such as *Tanin*, *Türk Yurdu*, and *İkdam*. In addition, in 1918 the popular journal *Yeni Mecmua* (New Magazine) published a special issue entitled 'Gallipoli: 5–18 March 1331–1915, Yeni Mecmua's Extraordinary Issue'.[10] It dedicated the whole issue (approximately 150 pages) to the Gallipoli campaign. Similarly, in 1918, the *Donanma Mecmuası* (Magazine of the Navy), published

'Gallipoli 5–18 March Victory' as a special issue which aimed to 'honour the memory' of this victory.[11]

Thus, during the war, 18 March 1915 was celebrated as a great naval victory. After the foundation of the Turkish Republic, however, it took equal billing with victory in the 'Battle of Anafartalar' at Suvla—the MEF's final throw of the dice at Gallipoli in August 1915. The reason for this was the supposedly pivotal involvement of Mustafa Kemal, founder of the republic.

Remembering Gallipoli in the Turkish Republic

Defeated overall in 1918 and dismembered in the Treaty of Sèvres, from the ashes of empire emerged a proud and victorious nation, the Republic of Turkey. It was founded in 1923 after the War of Independence (1919–22). Although Mustafa Kemal had been a relatively minor figure in the CUP and a senior officer, subordinate to two or three others, at Gallipoli, he was undoubtedly *the* inspirational leader of the fight for independence and became the country's first president. He embarked on a rapid and ambitious programme of modernization, aimed at throwing off the decrepit Ottoman past. Modes of dress, of telling the time, of writing, and of talking were all reformed in a bid to sever links with the past.[12] This was both modernization and state-sponsored amnesia. Knowledge of the Armenian genocide was but one of many aspects of the Ottoman past that was left behind. Place names within Turkey (which had already been subject to change by the CUP from 1914, particularly Greek place names) were systematically changed from 1920 onwards: Kurdistan and Armenia were deemed not to exist; buildings which betrayed previous Armenian connections had their telling inscriptions erased. In 1924 the Turkish National Assembly abolished the caliphate, and in 1928 the constitutional clause identifying Islam as the religion of the Turkish Republic was struck out. Also in 1928, on 1 November, the Latin alphabet replaced the classical Arabic-based Ottoman script. Nor was Arabic to be used henceforth in education or in public; even the call to prayer was to be chanted in Turkish. In

1934, the Surname Law required everyone to adopt Turkish surnames that included no hint of other ethnic identities.[13]

Kemal's policies can be seen as a continuation of the Turkish nationalism of the CUP. In place of the multi-ethnic, multi-religious, and multilingual Ottoman culture which the CUP had already begun to attack, the Turkish Republic strove to define a homogenous, secular, and monolingual Turkish identity. This process was underpinned by what Erik Zürcher has described as 'ethnic cleansing on a massive scale' as well as by the loss of the empire's Arab provinces.[14] In demographic terms, the non-Muslim proportion of the population living within Turkey plummeted from 20 per cent pre-war to 2.5 per cent after the war.[15]

In tandem with the severance of these multifarious reminders of the Ottoman past, came a concerted project to rewrite its history. Unacceptable books such as those about or in the language of the Armenians or the Kurds were destroyed, and a new Turkish historiography was developed, particularly through the Association for the Study of Turkish History, established by Mustafa Kemal in 1930.[16] The origins of Turkey's culture were sought in the pre-Ottoman period, tracing Turkishness back to the Sumerians and Hittites of Anatolia. Only the most glorious episodes of the Ottoman Empire were celebrated, particularly the reigns of Sultan Mehmed II (Mehmed the Conqueror, r. 1444–6, 1451–81) and of Sultan Süleyman I (Süleyman the Magnificent, r. 1520–66). Kemal took the lead in this process through epic speeches, most notably in his thirty-six-hour long 'Nutuk' speech delivered to the Turkish Grand National Assembly in 1927, which established the origins of the Turkish nation in 1919 and his own role in initiating the War of Independence.[17] Interestingly, however, Kemal's own writings on the Gallipoli campaign, *Report on the Ariburnu Battles* and *A Short History of the Battles at Anafarta*, were only published posthumously in 1968 and 1943 respectively.[18] The memory of Gallipoli, it must be acknowledged, played only a minor role in building a new national identity.

In building a new history for the country, the most recent era of the Young Turks and especially the CUP's period in power after 1913 was dismissed and despised.[19] This had begun as soon as the CUP leadership fled the country on 1 November 1918. The political opposition in the Ottoman Parliament was now free to criticize the 'gangster administration' of the CUP government for all its failings in the war, and in particular to establish its responsibility for the Armenian massacres, with a view to forthcoming legal retribution.[20] Kemal did make one public reference to the massacres in a speech in December 1919 in which he denied its severity and blamed the victims for the violence against them.[21] But thereafter, public discourse and the new historiography were silent on the matter of the genocide—rather than attempting to explain it away, it was not mentioned at all.[22] Thus there was always a profound ambivalence about remembering the First World War in the Turkish Republic. Only 131 studies of the war were published in Turkish prior to 1955, a tiny fraction of the number published in Britain, France, or Germany.[23] For some, the war was inseparable from their criticisms of the CUP, but for others, 1914–18 paved the way for the War of Independence (1919–22) and the foundation of the Turkish nation state. Meanwhile, the memory of Gallipoli stood apart from such debates and insofar as it was remembered, it was increasingly known for the role of Mustafa Kemal as well as for the devotion of the country's ordinary soldiers.

1925–1945: Kemal, Mehmets, and Johnnies

The first two decades of the Turkish Republic constitute the initial high-water mark in the celebration of Mustafa Kemal's role at Gallipoli. Largely unknown during the First World War, either in the Ottoman Empire or beyond, after the foundation of the republic he was frequently referred to as 'the hero of Anafartalar', 'the man who changed the destiny of his country', and 'the father of the Turkish nation', an idea encapsulated in the name Atatürk. It was a myth that was as important for the creation of the Turkish nation state as it was useful for his erstwhile enemies: in his 1923 account of the campaign, Winston Churchill made much of the inspirational and dynamic leadership of Kemal.[24] Indeed, Ayhan Aktar argues that the British were instrumental in creating the legend of Kemal's role at Gallipoli. It was Churchill who coined the phrase 'Man of Destiny', a phrase Cecil Aspinall Oglander was encouraged to repeat in his British official history of the campaign for useful diplomatic effect.[25] This idea then reverberated in Turkey such that the Gallipoli campaign (referred to as 'Çanakkale' in Turkey) became synonymous with him. As the *Cumhuriyet* newspaper put it in 1936: 'Çanakkale means Atatürk.'[26] His dramatic order during the landing of the Anzacs on the Gallipoli peninsula became nigh-on ubiquitous whenever the campaign was remembered: 'I do not expect you to attack, I order you to die! In the time which passes until we die, other troops and commanders can take our place!'[27]

Until his death in 1938, it became the custom to send Atatürk greetings via telegram from Çanakkale during commemorative events as a means to express gratitude and reverence for his military leadership during the campaign.[28] From at least 1930 onwards, these events were mostly organized by the Committee for Cemetery Restoration (Şehitlikleri İmar Cemiyeti).[29] The committee hired a boat called the *Gülcemal*, and from early July advertisements in Turkish newspapers advised that tickets were available for purchase at the Sirkeci train station in Istanbul (the newly formalized name for Constantinople under the republic). Interestingly, the commemorative trips were not confined to a specific day or even a specific month in the early 1930s, and ranged instead from July to late August or even early September. Moreover, until 1933, the commemoration events took place on board without landing; the *Gülcemal* boat made its way up the coast towards Ari Burnu and Anafartalar (i.e. Anzac and Suvla) and the anniversary was celebrated with the national anthem, fireworks, Mawlid (the celebration of the Prophet's birthday), and lectures on the importance of the Gallipoli campaign on the boat itself. Only from 1933 were the ceremonies conducted on land.[30]

What had been happening on the Gallipoli peninsula up until that point? After the evacuations by Allied forces in December 1915 and January 1916, Ottoman forces occupied the peninsula until the end of hostilities on 31 October 1918. Thereafter British forces took control, a range of foreign forces returned to Gallipoli, and considerable numbers of aimless Turkish soldiers remained behind. While the civilian population set about re-establishing their livelihoods on the peninsula, Australian official historian Charles Bean returned to gather evidence and museum relics with the Australian Historical Mission.[31] A Graves Registration Unit also began its work shortly after the armistice, continuing through much of 1919 before the Imperial War Graves Commission (IWGC) took over.[32] The right to build cemeteries and memorials on the former battlefields was enshrined in the peace treaty.[33] It took five years and thousands of Turkish, Greek, Italian, Russian, and Armenian workers to construct them.[34] From November 1920 there were additional residents on the peninsula: the remnants of the Russian White Army, their wives and children—some 26,595 individuals in total—who were interned near the town of Gallipoli (and also on the island of Lemnos and at Chataldja).[35] There was also a small French garrison nearby, as well as 15,000 British troops who were protecting the peninsula. The presence of personnel from the British Empire on the peninsula meant that from 1919 onwards, for several years at least, Anzac Day was marked back at the landing site. Newspaper sources suggest thirty-four people were present at the service in 1920,[36] but photographs of the service in 1923 show several hundred were in attendance.

From August 1920, Gallipoli had been part of the international zone of the straits designated by the Treaty of Sèvres. Under the treaty, Gallipoli was ceded to Greece, but its burial grounds became British. Its disputed status nearly sparked further belligerence between Britain and Turkish nationalists during what was termed the Chanak crisis in September 1922.[37] After the Treaty of Lausanne, which replaced the Treaty of Sèvres in 1923, the cemeteries were placed in the hands of the IWGC.

Fig. 23. Three British priests, dressed in white, read prayers for the fallen at the 1923 Anzac Day ceremony at Anzac Cove.

Fig. 24. Another view of Anzac Day at Anzac Cove in 1923. The relatively small congregation of military personnel is in stark contrast to 21st century attendance at commemorations on the peninsula.

Although some intrepid individuals visited the peninsula independently, it was only from 1925 and the completion of construction work that organized pilgrimages to the cemeteries began. In addition to the *Gülcemal* visits, there were at least seven organized pilgrimages from ex-Allied countries prior to the Second World War, and in addition cruise ships that toured the region would regularly land their passengers at Helles to visit the battlefields.[38] These pilgrimages were sometimes organized as part of the unveiling of a new memorial. The only exclusively Australian pilgrimage between the wars was in 1929. Otherwise, such trips departed from a UK port, and were available only to those wealthy enough to make the journey from Australia, New Zealand, or the UK out to the eastern Mediterranean.

One of these pilgrimages prompted a momentous exchange. In 1934, a 750-strong group of veterans and bereaved relatives travelled to Gallipoli from Liverpool on the *Duchess of Richmond*, arriving on 30 April. Five days earlier, the newspaper *Cumhuriyet* (Republic) published a short section of a long letter written by an Anzac soldier to Atatürk. The letter praised the Turkish soldier as a 'decent, brave and generous enemy'.[39] On the same day, Atatürk sent a message in response to a Melbourne newspaper's request ahead of Anzac Day. It was briefly reported in a number of Australian newspapers and reprinted a day later in *The Times* under the headline, 'Anzac Day in Australia: The Ghazi's Message':

> The Gallipoli landing and fighting on the Peninsula showed to the world the heroism of all who shed their blood there, and how heartrending for their nations were the losses this struggle caused.[40]

It is worth noting the relative obscurity of the reporting of these sentiments, given the later fame of the speech from which they are drawn. The full message conveying and expanding on these sentiments was read out by Şükrü Kaya, Atatürk's interior minister, beside the Mehmetçik memorial.[41] The speech was not referred to in further reports on the pilgrimage, which only landed at Gallipoli some days after 25 April.[42] That message was nonetheless extraordinary:

> Those heroes that shed their blood and lost their lives are now lying in the soil of a friendly country. Therefore rest in peace. There is no difference between the Johnnies and the Mehmets to us where they lie side by side here in this country of ours. You, the mothers, who sent their sons from far away countries, wipe away your tears. Your sons are now lying in our bosom and are in peace. After having lost their lives on this land, they have become our sons as well.

These words were written within twenty years of the campaign, in a generous and magnanimous spirit towards men who might otherwise have been seen as imperialist invaders. Why? The answer lies, perhaps, in those invaders' inadvertent role in creating the Turkish nation. With the project to rewrite Turkish history in full swing, and Kemal's heroic role as Atatürk (father of the Turks) in the new republic established in 1923, he held a secure vantage point from which to view the virtues of the rallying power of a defensive victory such as Gallipoli in comparison to the disastrous aggressive, expansionist campaigns waged by Enver Pasha. As Adrian Jones has argued, 'Turkey was born there and then at Gallipoli, as Johnnies who knew nothing about Mehmets re-made those Mehmets as masters of their fate.' Hence Kemal's ability to appreciate their heroism.[43]

Despite it being addressed to an audience of ex-servicemen who were presumably predominantly British, this message is representative of the early republican habit of conceiving of the enemy at Gallipoli as the Anzacs (as opposed to the British and French, who were emphasized in the late Ottoman period). Interestingly, Atatürk's 1934 message notwithstanding, such references became increasingly respectful and admiring during the Second World War. Turkey was neutral during the war but had made various commitments to Britain via the 1939 Treaty of Mutual Assistance. The arrival of the 2nd Australian Imperial Force (AIF) in Egypt prompted this reflection on their predecessors in *Cumhuriyet* by the prominent columnist Abidin Daver:

Anzacs are terrific soldiers... in Çanakkale, the Anzacs showed themselves worthy opponents for the brave Mehmetçiks. They successfully ascended the hard, steep slopes Liman von Sanders had left unguarded in the expectation that landing there would be impossible. Their fierce and heroic attack could only be stopped by Lieutenant Colonel Mustafa Kemal and the heroic Mehmetçiks. Now, the Anzacs have returned to Egypt and have been sent to unknown regions. This time their arrival in the Orient is friendly, not hostile. Though we wish for no war, should war break out, the Anzacs will no longer be face to face with Turks, but neck and neck with them in the race to heroism.[44]

This Phoney War-era comment celebrated the prowess of both generations of Anzacs, the better to affirm the fighting power of Turkey, then and now.

With Anzac Day well established in Australian culture, these soldiers marked the twenty-fifth anniversary of the campaign in Egypt. This was warmly reported in the Turkish media. Through their commemorations, the new generation of Australian soldiers 'were able to appreciate the value of the defenders of Turkish soil'. The Turkish correspondent thereby enthusiastically announced,

This history, filled with horrific memories of the Great War and with feelings of friendship and admiration towards the new ally Turkey, will now introduce both the Anzacs and the firm connection binding Turkey to its friendly and heroic sons to coming generations.[45]

Despite such reports, by contrast, Turkish public commemorations of Gallipoli appear to have come almost to a complete halt during the war, or at the very least were not widely reported. After all, they would have served as a reminder of both President İsmet İnönü's non-involvement in Gallipoli and of the earlier alliance with Germany in an unhelpful manner given the Second World War stance of non-belligerency. Commemorations did resume at the war's end, but took on a fearful note amidst the uncertain climate in foreign relations. Thus one report in 1946 warned that the 'events of 18 March could be repeated any time and Turkey should be careful'.[46] Indeed, the influence of international relations upon the commemoration of Gallipoli is particularly noticeable in the early post-war decades.

The 1950s: Building Friendships and a Memorial

The one-party system in Turkey ended in 1946, and the first free elections in 1950 brought victory for the opposition Democrat Party. Turkey had already aligned with the West in the Cold War, received aid under the Marshall Plan from mid-1948, and under the Democrat Party embraced free-market economics encouraging further foreign investment. This was a major departure from the neutral stance of Kemalism.[47]

The realignment culminated in Turkey's participation in the Korean War with American and United Nations' forces, fighting alongside its erstwhile enemies: Britain, France, Australia, and New Zealand. A variety of warm words and gestures ensued—this time encompassing not just the Anzacs, but Britain and France too. Thus, proud of the latest chapter of their country's military history, and seemingly keen to create a physical bond between the two campaigns, Turkish soldiers in Korea sent home a national flag via the Association of Turkish Students to be presented at the 18 March ceremony in 1952.[48] In return, a handful of soil from the veterans' cemetery in Gallipoli was sent to Korea.[49] The following year, Milliyet newspaper wrote in its Çanakkale week coverage (i.e. commemoration of 18 March) about 'our present very valuable allies and beloved friends [the] English and French naval and army forces...we laid the foundations of our today's friendship and alliance 38 years ago in [the] Çanakkale and Gallipoli battles'.[50]

Renewed military action in Korea seems to have re-energized the commemoration of earlier warfare. The 1952 preparations for the August commemorations of the Battle of Anafartalar were extensively covered, and prompted numerous veterans from across the country to write to express their gratitude.[51] The actual commemorations included a pilgrimage/visit by a 420-strong contingent on the boat *Etrüsk*, which departed from Istanbul. It included 'representative groups of the veterans from England and France' and also more than thirty retired generals, veterans, the families of the war martyrs, national and international media, the members of the Çanakkale Memorial Committee, and youth representatives.[52] There was a military ceremony before their departure, and while the party visited the battlefields, military aircraft flew overhead and dropped flowers as a means to express their respect to the martyrs.[53] In addition, there were other lower-key demonstrations of friendship at Anzac Day events and similar occasions in this period that were used as soft diplomacy to cement the new NATO alliance.[54] These included the attendance of Turkish representatives overseas on Anzac Day 1953 for the first time. In August that year, it was particularly emphasized in newspaper coverage that the Turkish commemorative committee in Gallipoli would be visiting British and French memorials on the peninsula to lay wreaths.[55]

Nationalist university student groups appear to have been particularly enthusiastic about commemorating Gallipoli. Contingents from both the Federation of Turkish Students (Türkiye Talebe Federasyonu) and the National Association of Turkish Students (Milli Türk Talebe Birliği, MTT) travelled to Çanakkale in March 1952,[56] and an exceptional number of university students, some 500 in total, attended the ceremony.[57] Later in the year, the National Association of Turkish Students initiated the bid to build a monument for the Çanakkale martyrs,[58] while in 1953 the association organized 'Çanakkale week' in Eminönü (the historical heart of Istanbul). Such enthusiasm lasted well beyond the generation of students of the Korean War era. The Eminönü commemorations saw their biggest crowd in fifteen years (its exact size unspecified) in 1958.[59] In that year, university students from Çanakkale began to campaign for 18 March to be made a national holiday: they surveyed 25,000 people (20,000 respondents were said to view their demands positively) and petitioned the prime minister with their idea.[60]

Students were also considered to form the greater part of the approximately 1,000-strong crowd at the commemorative events in Çanakkale in 1959, which featured a traditional-style Ottoman military band (Mehter Takımı).[61]

In contrast to these significant but not extraordinary crowds, the forty-fifth anniversary of the Çanakkale victory in 1960 was attended by 10,000 people from Çanakkale who were joined by thousands from other cities,[62] in advance of the opening of the towering Çanakkale Martyrs' Memorial later that year.[63] Before then the focus of Turkish ceremonies on the peninsula had been the modestly proportioned Sergeant Mehmet Memorial (Mehmed Çavuş Anıtı) at the Nek, scene of a desperate Australian attack on 7 August 1915. The only other Ottoman/Turkish memorial on the peninsula was reputed to have been destroyed by the Greeks during their occupation of the peninsula under the Treaty of Lausanne.[64] By contrast the Çanakkale Martyrs' Memorial was to be 135 feet tall and positioned at Eskihisarlık Burnu to overlook the Narrows. There was a groundbreaking ceremony held on 17 April 1954.[65] By 1957, 700,000 lire had been collected, but a further

Fig. 25. Sergeant Mehmet's Memorial is situated at the Nek. Its base dates from 1919 and the obelisk was added later. This was the only Turkish memorial on the peninsula prior to 1960.

1 million lire was required for the ambitious scheme.[66] Reports on the progress of the growing height of the memorial as it was constructed and on the progress of the fundraising campaign were published frequently during 1958. Among the donors were New Zealand veterans and, in the letter accompanying the donation, the New Zealand prime minister wrote to his Turkish counterpart, Adnan Menderes, saying that 'In Korea Turkish soldiers displayed the same heroism and solidity as they had done in Gallipoli'.[67] In return, the Turkish government acquiesced to a request from veterans in Ashburn, New Zealand for some stones from Gallipoli with which to build a memorial of their own.[68] By 1958, as the fundraising campaign continued, the absence of Turkish memorials on the peninsula was forcibly pointed out in *Milliyet*. Photographs of the British and French cemeteries were captioned 'No Turkish cemetery in Çanakkale', while two further photographs contrasted 'The Memorial erected for the defeated British soldiers (height: 32 metres)' with 'The Stone erected for the victorious Mehmetçik'. Further commentary on the matter adjacent to the photographs read, 'Çanakkale is full of memorials, yet these memorials have been erected for British and French soldiers who were defeated by the heroic Mehmetçik'.[69] The Martyrs' Memorial was completed in 1960, but there was dissatisfaction expressed about the absence of appropriate landscaping around it: the site was muddy, treeless, and somewhat barren in appearance.[70] In 1963, 500,000 saplings were set aside for the memorial.[71] It had to have its roof repaired in early 1966.[72] Nonetheless, it belatedly makes an important point. Henceforth, the dominant memorial on Turkish soil marked Turkish victory.[73]

1960–1980: A Unifying Example in Troubled Times

The crisis which erupted in Cyprus in 1964 and led to the 1974 Turkish invasion and ultimately the partition of the island, demonstrated the present-mindedness of commemoration. Just as the August 1960 opening of the Martyrs' Memorial had been helpful in bolstering the position of the new military junta that had taken power in a coup in

May that year, the dominant rhetoric of the time continued to draw on Gallipoli as a 'timeless' victory for the Turkish army that could be repeated whenever the country was under threat. Thus, for example, the secretary general of the Turkish National Students' Union (Türk Milli Talebe Federasyonu, TMTF), Nafiz Duru, warned the Cyprus government during the forty-ninth anniversary of the Gallipoli campaign: 'As history is repeating itself, we want to repeat the Çanakkale Victory once more in Famagusta.'[74]

This was a period of considerable instability in Turkey. The relatively liberal constitution of 1961 (prepared after the military *coup d'état* on 27 May 1960) allowed wider sections of society to publicly criticize Turkey's capitalist economy and its alliance with the USA and especially with NATO. Just as elsewhere in the world, there were vivid intellectual debates, student debating societies (*fikir kulüpleri*), and newly emergent left-wing journals such as *Devrim* (Revolution), *Yön* (Direction), and *Aydınlık* (Enlightenment) at this time. Following the intervention of the International Monetary Fund (IMF) in Turkey in 1958, foreign and particularly American investment had grown, and Turkey's economic relations became a source of controversy. In the late 1960s there were a series of mass demonstrations and violent clashes between left-wing and right-wing groups, with the police tending to side with the latter. When students dared to interrupt Gallipoli commemorations in 1967, they were rebuked by the minister of transport, Seyfi Öztürk, who invoked the unifying example of its memory: 'Let no discord arise. We will not find a way to unite unless the Çanakkale martyrs unite us. That is what the martyrs command.'[75] At the following year's commemorative cemetery, Öztürk returned to his theme and implicitly criticized the representative of the National Organization of Turkish Youth (Türkiye Milli Gençlik Teşkilatı) for politicizing the Gallipoli ceremony: 'Here is not a political rally ground.'[76] As political violence and unrest grew in Turkey, a second *coup d'état* in 1971 was followed by weak coalition governments, tremendous economic crises, and severe political violence with political arrests, allegations of torture, and perhaps 5,000 casualties between

1976 and 1980.[77] Earlier rebukes notwithstanding, Gallipoli became a platform, particularly for students, to demonstrate against the political climate. In 1974, for example, the nationalist right-wing organization, Ülkü Ocakları, used the Çanakkale commemoration as a platform to demonstrate against the government.[78] With ongoing disputes between ultra-nationalists and Islamists, students were ultimately barred from attending Gallipoli commemorations in 1976 and 1977.[79] Meanwhile, government politicians continued to claim Gallipoli's legacy for their own purposes, for example, extolling its soldiers as exemplars of loyalty in economically difficult times as in President Fahri Korutürk's 1979 message: 'This victory is the declaration of what can be achieved even in poverty by the Turkish soldier who is the symbol of courage and patriotism.'[80]

A third *coup d'état* in September 1980, led by General Kenan Evren, resulted in military rule for three years in a bid to restore stability and Kemalist principles. Although new political parties were allowed from 1983, and a civilian government was elected at that point, Evren remained president until 1989. Under military rule, with civil liberties curtailed and the abolition of political parties, Atatürk's reforms, his secularism and nationalism gained renewed prominence, as did the memory of the man himself.[81] The centenary of his birth was celebrated in 1981 with a publicity campaign: 'Atatürk is one hundred years old' (*Atatürk Yüz Yaşında*).[82] In March that year, President Evren sent a message to be read at the commemorations at Gallipoli; in August, he attended in person. Staying in Çanakkale for three days, he visited veterans' cemeteries and the battlefields during the Anafartalar anniversary. Evren gave a speech celebrating Atatürk as the 'hero of Anafartalar', vowing to defend him from any criticism of his role in the campaign.[83] During Çanakkale week the following year, Evren used Gallipoli as the means to declare the army's role in protecting the nation:

Today the honourable Turkish nation is more cautious and alert than ever before. It is tied to the Armed Forces with great faith and trust and is in absolute unity with it. From wherever the danger may appear, it remains mighty enough to meet it immediately.[84]

The veneration of Kemal continued throughout the decade. Evren's speech at the 18 March commemorations in 1987 encapsulates the most common messages about the important lessons taken from this defensive victory:

> The Çanakkale Battles, which have an unforgettable place in Turkish war history and represent the phrase 'Çanakkale is impassable' in world war history, showed the power of Turks and their invincibility in the most striking way and won for our country a unique soldier and genius like Atatürk.[85]

Some measure of the impact of these speeches and campaigns was revealed when a TV programme about the 'Çanakkale Martyrs' was broadcast on 18 March that year, and generated a large number of complaints to the broadcaster after it failed to mention Atatürk.[86] One of the most important themes that were elaborated when Atatürk and Gallipoli were discussed was the supposed direct link that could be drawn between the 1915 campaign and the republic: in 1986, the interior minister Yıldırım Akbulut said that the 'Çanakkale Victory led to the first steps being taken towards the foundation of the republic'.[87] Two years later in 1988, during the seventy-third anniversary of the campaign, the minister of justice Oltan Sungurlu said: 'Atatürk entered into the hearts of the Turkish nation and earned its trust, and with this trust he became the leader of the Turkish Republic.'[88] In such ways the two parts of Kemal's biography, which were relatively tenuously connected, were woven together into a renewal of the foundation myths of the republic.

Yet in addition to this promotion of Atatürk, Evren used Islam as a means to unite the country, in what was termed a 'Turkish-Islamic synthesis'. He quoted from the Quran and the Hadiths (sayings of Prophet Muhammad) to encourage school attendance, and decreed that 'The Knowledge of the Culture of Religion' should become part of the basic curriculum of every school. However, this was done in such a way as to be compatible with the secularist ideals of Kemalism. As Erik Zürcher has explained, 'this religious teaching was exclusively Sunni in content, and patriotism and love for parents, the state and the army (in the shape of the Turkish "Tommy Atkins"'—*Mehmetçik*) was presented as a religious duty'.[89]

It follows, therefore, that from the time of the military government onwards, the concept of martyrs became even more prominent in the commemoration of Gallipoli, notwithstanding its currency from the very start. The powerful symbolism of the martyr entails the religious duty to love one's country and sacrifice one's life for its unity. In contrast to the English equivalents of 'The Fallen' or 'The Glorious Dead', 'martyrs' has distinct religious overtones: Şehit (Shahid in Arabic) means those who died as martyrs for the faith. In the context of Turkish nationalism, however, the term şehit is used to refer to the soldiers who died as martyrs for the very existence of the state; in this highly militarized rhetoric, it is 'a sacred ideal' to sacrifice one's life, as Jenny White observes, for 'the unity and integrity of the state'.[90]

Remembering Gallipoli since the 1990s: Stronger and Broader

Three trends are noticeable in the way Turkey has remembered Gallipoli since 1990 or thereabouts. Firstly, the use of Gallipoli to build friendly international relations has been actively pursued and has achieved new levels of intensity. Secondly, interest in Gallipoli within Turkey has developed strongly. Thirdly, reflecting political developments within Turkey, the rhetoric used relating to Gallipoli has shown profound changes in the way Turkey has used its commemorations and has drawn on its past to define its national identity.

The friendly tone of relations between former enemies has always been apparent—it has already been noted in Kemal's 1934 address, and it was reflected in 1971 when Queen Elizabeth II and Prince Philip visited the Çanakkale Martyrs' Memorial and the new war museum in its ground floor. The queen wrote in the comments book: 'This is a memorial which does justice to the martyrs of the Gallipoli Battles in which the two nations gained a mutual, perpetual respect for each other.'[91] In recent decades, warm words have become the basis for what seems to be a diplomatic strategy on the part of Turkey and Australia. This has developed in tandem with growing popular participation in the commemoration of the campaign in both countries.

The first manifestation of these diplomatic gestures came in 1985, and included the unveiling of a monument bearing Kemal's 1934 address at the renamed Anzac Cove, as well as in Wellington, Canberra, and Albany (Western Australia, the departure port for the Anzacs). This monument should be seen in the context of a spate of Turkish activity on the peninsula: during the 1980s more than twenty monuments were built there.[92] Then, in 1990, the seventy-fifth anniversary of Anzac Day was attended by international political leaders and—at substantial cost to the Australian government—a delegation of

Fig. 26. A historical Ottoman band participates in the Turkish commemorations at the Çanakkale Martyrs' Memorial in 2007. The relief on the memorial behind shows an image of reconciliation between Ottoman and Anzac soldiers.

Australian veterans. The commemorations were an opportunity for high-level political meetings. The friendship between Turkey and her 'old enemies' was a repeated theme.

The veneration and care lavished upon Australian veterans inspired some Turkish veterans to highlight their plight after the ceremonies. A series of news articles appeared about their financial difficulties.[93] In November 1990, the surviving veterans were awarded a medal of honour: finally the rhetorical veneration of Mehmetçik was being acted upon.[94] Thereafter, the memory of Gallipoli became increasingly prominent in Turkish public life. It is part of a broader interest on the part of Turkish citizens in a past that the republic had tried to erase.[95] In 1992 the Çanakkale Onsekiz Mart University—a university named after the victory—was founded, which has done a great deal of work to develop scholarship and popular engagement with the commemoration of Gallipoli. Another indicator of the heightened awareness of the campaign in the 1990s was the launch of a new 500,000 Turkish lira note on 18 March 1993 which featured the image of the Çanakkale Memorial on its reverse. One year later, a gold medal was awarded to the city of Çanakkale.[96]

At the same time as these developments, a crisis in Turkey's national identity and territorial integrity emerged. During the 1990s, identity politics came to the fore in Turkey, and Sunni Muslims, Kurds, and Alevis became more prominent after decades of Kemalist emphasis on a mono-ethnic state. An armed Kurdish movement, led by the Kurdistan Workers' Party (PKK) had been founded by Abdullah Öcalan in 1978 in order to 'establish a socialist Kurdish state in the southeast of Turkey'.[97] It was in the 1990s that the movement reached its peak, engaging in acts of sabotage, kidnapping, and bomb attacks in a bid to promote their cause. In 1999 the capture of Öcalan led to the neutering of the threat posed by the PKK.

In response to these activities, and despite a revolving door of ten coalition governments between 1991 and 2002, there was a remarkable rhetorical consistency from succeeding Turkish prime ministers, who often took the commemoration of Gallipoli as an opportunity to restate the country's territorial integrity. Thus, during Çanakkale week in 1993, developing the familiar theme of 'Çanakkale is impassable' and its message to would-be invaders (or in this case to internal threats), Prime Minister Süleyman Demirel announced that 'no single stone of the country can be passed'.[98] Four years later, during the eighty-second anniversary, Deputy Prime Minister Tansu Çiller said, 'What makes this nation a grand nation is the power to sacrifice one's life in order to make this soil the fatherland.'[99] Çiller's message was reiterated in President Süleyman Demirel's speech in the same ceremony:

> Historical turns like the 18th of March remind us how we have come to these days. The Turkish nation will never forget its history filled with grand epics, and will forever protect the unity and solidarity of our country and its indivisible integrity.[100]

Amidst heightened tensions, in an extraordinary lapse of military discipline, soldiers and their officers shouted slogans during the Çanakkale commemorations of 1998, like 'The Mehmetçik is here, where is the PKK? Martyrs never die and the fatherland will never be divided.'[101] The eternal example of the heroic Turkish soldier from Gallipoli was thus contrasted with the PKK, as a reminder of the nationalist commitment to the indivisibility of the nation, demonstrated here by aggressive language and a highly visible military presence.

Since the resolution of the threat from the PKK, however, the door has opened to a significant change in the way in which the First World War is remembered in Turkey. The Refah Party, which built from local government election victories in March 1994 to win power at the national level in 1996, began to put forward a revised narrative of Gallipoli, which fitted with their Islamist and Ottomanist viewpoint— that is, an alternative to the Kemalist and Turkish nationalist idea of the campaign's significance. Where local government was controlled by the Refah Party, massive tours to the peninsula were organized to promote this view.[102] Thus change was under way before the rise to power of the Justice and Development Party (Adalet ve Kalkınma Partisi, AKP), the Republic of Turkey's first governing party to be strongly disassociated from Turkey's military. The AKP came to power in 2002 and Recep Tayyip Erdoğan became prime minister in 2003. Although the party has Islamic roots, it has presented itself as an economically liberal but socially conservative party of the Right. Given the novelty of its strong stance against the military, some commentators have suggested that the AKP is attempting to abolish Kemalism and the cult of Atatürk. Its actions in relation to Turkish and Ottoman history have therefore been closely watched.

The most significant change is that Erdoğan and his government have set out to commemorate the First World War as a whole, and not just Gallipoli, as in the past. There are sound reasons for doing so from a historian's point of view, insofar as the Ottoman Empire's part in the war has been significantly underrepresented in the international historiography of the war. In the context of the centenary years, it is important to represent the full breadth of the war's impact on the empire and vice versa. But there are further incentives for this shift that are specific to the AKP and Turkish politics. Interest in Ottoman history has grown over the last three decades, but particularly for the AKP, the glorious and victorious Ottoman past is recalled to make Turkey a regional power in the Middle East.[103] Furthermore, the Ottoman past offers 'an alternative source of shared identity',[104] and it was no longer felt that the last four years of the empire needed to be avoided, or that the culpability of the CUP for the disasters of the First World War had to be criticized. We can observe this process with regard to the treatment of Enver Pasha and the Battle of Sarıkamış.

The rehabilitation of Enver Pasha predates the AKP government. In 1996 his remains were brought to Turkey from near Ab-i Derya and reburied at Abide-i Hürriyet (Monument of Liberty) cemetery at Şişli, Istanbul. The minister of war during the First World War, who had been either ignored in Turkish official history or at worst seen as responsible for the defeat of the Ottoman Empire given his decision to ally with Germany, was emphatically praised for the first time by a

Turkish president in July 1996, when Süleyman Demirel explicitly said, 'Enver Pasha is a hero'.[105]

One of Enver's worst defeats of the First World War was Sarıkamış. It came immediately before Gallipoli and, indirectly, was its cause in that it prompted the appeal to Britain for assistance from the Grand Duke Nicholas, commander of the Russian forces, which led Britain to launch the attack on the Dardanelles. At Sarıkamış, high in the Caucasus Mountains, Ottoman soldiers fought the Russians in December 1914 and January 1915. Amidst freezing conditions, Enver's men had to march through deep snow, since the front was 250 miles beyond rail communications. The sub-zero temperatures (minus 31 degrees centigrade at Ardahan, for example)[106] claimed far more lives than did the Russians: 90,000 of the 150,000-strong Ottoman force died.[107]

Since the debacles of the CUP's wartime leadership were forgotten for decades in the Turkish Republic, the first time that Sarıkamış was commemorated was in 2003.[108] By 2010, Erdoğan was encouraging his countrymen to remember not just Gallipoli but more difficult souvenirs. His 18 March speech that year included this passage:

> Do not abandon or orphan this martyr's cemetery at Çanakkale, or those cemeteries at Sakarya, Dumlupınar, Sarıkamış and all the others which we have rebuilt from top to bottom, and to which we have given an entirely new appearance.[109]

Continuing this developing engagement with Sarıkamış, 2013 saw not just an award-winning feature film on the subject,[110] but considerable popular interest in its commemoration. Ten thousand people from across Turkey participated in a commemorative 5-mile march for the ninety-eighth anniversary of Sarıkamış. The event was promoted by the local governor Eyüp Tepe and the youth and sports minister Suat Kılıç. The local people interviewed emphasized that they lived together as Turks and Kurds. The journalist reporting this event concluded his article, 'Commemorating the martyrs eliminates the ethnic and political divisions and unifies the society.'[111]

Indeed, the other notable change in the memory of the First World War developed by the AKP is the reintegration of the Kurds into the prehistory of the Turkish Republic. In conjunction with pressure from a range of groups from Turkish civil society, there have been important moves towards a peaceful resolution of the Kurdish issue, including the legalization of the use of the Kurdish language in 2001,[112] and the setting up of a Kurdish-language TV station in 2009.[113] Part of this project, according to the academic Cengiz Güneş, has been the AKP's emphasis on Turkey's Islamic heritage, with the aim of depoliticizing the Kurdish identity. In that sense, the First World War in general and the Gallipoli front in particular have been interpreted and recreated as historical moments where the idea of a supra-identity above ethnic differences was possible. Thus Erdoğan included Kurds in his Gallipoli rhetoric, such as his speech during the commemoration ceremony on 18 March 2013 when he emphasized that the victory in Gallipoli was not the result of one single ethnic group (i.e. the Turks) but a collective achievement.[114] Others followed his lead, such as the governor of Diyarbakır who, two months after Erdoğan's speech, promoted a similar message through a bilingual billboard written in Kurdish and Turkish. It said: 'in Gallipoli, there is also a piece of you!'[115] Bearing in mind the deportation of Kurds from Diyarbakır province and elsewhere from 1916 onwards—a policy aimed at the Turkification of this and other Muslim Ottoman regions[116] (albeit a policy shorn of the deliberately murderous treatment of the Armenians)—this is a remarkable acknowledgement of the Kurdish part in Ottoman history.

This encouragement to remember the full range of sacrifices of the First World War can be read as an attempt to challenge the traditional nationalist history of the republic. But in using a shared Islamic heritage as a means to build a shared identity, it excludes other religious minorities from being remembered as part of the prehistory of the republic. In particular, the new memory of 1914–18 in Turkey has one blind spot: the Armenian genocide.

A brief outline of what happened to the Armenians is necessary. A substantial Armenian population, a Christian minority, resided in eastern Anatolia, but was also scattered throughout the Ottoman Empire. In early 1915, a developing pattern of violence against them emerged. In February 1915, prompted by the defeat at Sarıkamış,[117] Armenians in the Ottoman Army had been disarmed, and some of their counterparts in labour battalions had been executed in March. Able-bodied men were thus targeted ahead of a more widespread onslaught against the population. Meanwhile deportations of significant numbers of Armenians in southern Anatolia had begun. Fearing similar treatment, Armenians in Van rose up against Ottoman forces from 20 April. They held off the threat initially, but triggered the wider onset of deportations against their brethren from 24 April onwards, with a further intensification of these activities from July to include western Anatolia. Many Armenians were killed immediately; others were robbed, raped, or murdered during the forced marches; or died from exhaustion, starvation, or disease as they were driven towards the desert regions of the interior. At least 800,000 Armenians died. Despite grave deficiencies in the historical record—important documents have been destroyed or remain inaccessible—a substantial body of scholars agree that these deportations and massacres were intentionally genocidal. In March 1915, the central committee of the CUP had empowered Dr Bahaeddin Şakir to 'eliminate the internal danger'.[118] It was Şakir's 'Special Organization' that enacted much of the violence.

The exact triggers for this sustained and brutal outbreak of violence against the Armenians are hotly contested. It has been argued—often by those who also downplay the severity of the violence and its genocidal intent—that the Armenians brought it upon themselves through their disloyalty, and that the Russians had been aiming to dismember the Ottoman Empire and had instigated an Armenian uprising as a means to grab traditionally Armenian lands within Ottoman territory. The argument then follows that the violence was a direct response to this threat. Often it is stressed that these massacres occurred in a generally lawless period in the final years of the empire, and that Turks, Greeks, Kurds, and others were also subject to sustained violence. These are the explanations that have been taught in Turkey since the early years of the republic.[119]

Meanwhile, others have looked to a more complex assessment of the response to the humiliation of a century or more of being written off as the sick man of Europe, and the deep fears of dismemberment at the hands of the encircling Great Powers. In this context, the early events of the First World War seemed to herald the imminent fall of the empire, and given the seeming inability to combat the enemy without, unleashed an almost primal level of violence aimed at the perceived enemy within. As the historian Taner Akçam explains:

> It was not a coincidence that the Armenian genocide took place soon after the Sarikamiş disaster and was contemporaneous with the empire's struggle at Gallipoli. As a rule, the acceleration of the process of a country's decline and partition helps to strengthen a sense of desperation and 'fighting with one's back to the wall.' As the situation becomes increasingly hopeless, those who have failed to prevent the collapse become more hostile and aggressive. When the crisis deepens, they resort to increasingly barbaric means, and come to believe 'that only an absolute lack of mercy would allow one to avoid this loss of power and honor.' A nation that feels itself on the verge of destruction will not hesitate to destroy another group it holds responsible for its situation.[120]

It therefore follows that although the cessation of the naval attack at the Dardanelles on 18 March came to be celebrated as a victory in Turkey, this was not immediately apparent. The Allied onslaught continued throughout much of 1915, as did the fear and panic engendered in the government. This informed the genocide. The link between the two campaigns is indirect, but it exists and should be acknowledged more widely.

The Armenian genocide has increasingly become a subject of study both inside and outside Turkey in the last ten years. This has been highly controversial. When the AKP came to power in 2002, there had been hopeful signs that the party was attempting to reconsider its relationship with Armenia. In 2006, Erdoğan declared that officials should replace the phrase 'so-called Armenian genocide' with the more neutral phrase '1915 events'. Two protocols were signed by Turkey and Armenia in 2009 with the aim of establishing diplomatic relations between the two countries. Moreover, on several occasions Erdoğan has suggested that the dispute over the history of the Armenians should be left to historians. An increasing number of Turkish scholars, such as Taner Akçam, Fatma Müge Göçek, and Fikret Adanır, have been writing on the tragic events of 1915. In 2005, the first conference on the issue of the Ottoman Armenians was held in Istanbul,[121] and publicly discussed the reliability of official Turkish state accounts of the Armenians during the war for the first time within Turkey. Other conferences have followed, and at the initiative of a coalition of human rights groups, the Armenian genocide has been commemorated in ceremonies held in Istanbul each 24 April since 2009. Some 3,000 people attended ceremonies held in six cities across Turkey in 2013.[122]

Yet the academic conference on the Ottoman Armenians was initially to have been held in May 2005, but had to be postponed when there were accusations in the Turkish Parliament that it amounted to treason against the Turkish nation. It was held that September with assistance from the AKP government.[123] Furthermore, academic discussion has been stifled by Article 301 of the Turkish Penal Code which criminalizes insults against 'Turkishness'. This has been used to incriminate numerous writers such as the Nobel Laureate Orhan Pamuk,[124] the novelist Elif Şafak, and the Turkish-Armenian journalist Hrant Dink. On 19 January 2007, Dink was assassinated by a 17-year-old Turkish nationalist shortly after the premiere of a documentary on modern world genocides in which he made comments about the Turkish denial of Armenian genocide of 1915.[125] Taner Akçam, who wrote an editorial in 2006 in defence of Dink and his work, went to the European Court of Human Rights to fight against Article 301. In examining the intimidation faced by him as a scholar working on the Armenian genocide, the case revealed numerous accusations made in the Turkish media that Akçam was a traitor and a spy in the pay of the German intelligence service, which had commissioned various of his scholarly works. Akçam's requests for published corrections, including defamation, were dismissed in a series of Turkish court cases. The European Court of Human Rights ultimately found in 2011 that Turkey's use of Article 301, for criminal prosecution of scholarship on the Armenian genocide on the grounds of denigrating Turkishness, amounted to a violation of freedom of expression.[126]

This is the context in which foreign interventions in the debate must be viewed. France, Russia, and nineteen other countries have recognized the massacres as genocide.[127] As Turkey sought to join the European Union, it came under pressure to acknowledge the genocide.[128] In March 2010, the US House Committee on Foreign Affairs recognized the genocide, and a full debate in Congress was only narrowly avoided after some intense diplomatic manoeuvres.[129] Erdoğan responded in his 18 March commemorative speech that year:

> I would like to underline that, just as this country's Mehmetçik is too great to fit into history, this country's history is too stainless, mighty, glorious and as true as the sun is bright to be contradicted by [other] parliaments. If they wish to clarify the events that took place in 1915 in Turkey's east, the address for this is not parliaments thousands of kilometres away, but the archives, documents, memoirs, reports, letters and the images. The irresponsible statements and unfair decisions we see today in some of the countries which, in a spirit of imperialism, spewed out death at Çanakkale and attacked the fatherland of the Turkish nation are nothing but a slander against a nation that is owed an apology... You cannot easily find in another nation's history the humanistic, prudent, and caring manner the Turkish nation has shown, even in times of war.[130]

These political gestures on the genocide by former imperialist powers were thus an affront to the great and benevolent nation of Mehmetçik.

The most recent parliamentary recognition brought the relationship between the Armenian genocide, Gallipoli, and diplomacy into even sharper focus. In May 2013 the New South Wales State Parliament recognized the Assyrian, Armenian, and Greek genocides perpetrated by the Ottoman Empire between 1915 and 1922.[131] This drew condemnation from the Turkish foreign minister Ahmet Davutoğlu, who drew on the sentiments of Kemal's 1934 speech:

> These persons who try to damage the spirit of Çanakkale/Gallipoli will also not have their place in the Çanakkale ceremonies where we commemorate our sons lying side by side in our soil.[132]

The New South Wales parliamentarians will not be given visas to enter Turkey. No criticism in this most sensitive of areas will be tolerated: warm relations come at the expense of candour. It is expected that an attempt to get Australia's Federal Parliament to recognise the genocide will not occur until *after* the Gallipoli centenary.[133]

Thus Turkey's relationship with Australia as 'a strategic partner',[134] which has been built on the back of the commemoration of Gallipoli, has an implicit quid pro quo wherein the Australian government will not tackle the issue of the Armenian genocide lest they jeopardize the centenary arrangements. The Turkish government has explicitly stated its aim to focus attention on only the Gallipoli part of the momentous events of 1915 during the centenary.[135]

Turkish interest in the Gallipoli campaign has grown strongly in the last decade. For years, the only Turkish-made movie about Çanakkale had been a 1964 vintage romantic tale of a Turkish soldier taken prisoner, who, forsaking a love affair with a British nurse, escapes to blow up a British facility.[136] By the time that Tolga Örnek made his documentary *Gallipoli: The Frontline* in 2005, interest was such that the documentary became the highest grossing in Turkish cinema history.[137] Furthermore, three new feature films about Gallipoli were released in 2012–13, and three more are being made.[138] In 2007, in a mirror image to the 5,000 backpackers at Anzac Cove, 8,000 Turkish youths marched in Bigali, a village elsewhere in Gallipoli, in a kind of re-enactment of the 57th Regiment's march into battle and its order to die under Mustafa Kemal in 1915.[139] This public interest was underpinned by infrastructure investment by the Turkish government—including the building of a controversial new road at Anzac Cove, and more recently one at Suvla. Thus whereas around the turn of the century, approximately 200,000 to 250,000 visitors attended Gallipoli's cemeteries, in 2013 Prime Minister Erdoğan boasted these figures had increased tenfold to between 2 and 2.5 million. Taking the credit, he noted: 'we did the physical arrangements that these cemeteries deserved and we will continue to do so'.[140]

Thus there have been two notable elements in Turkey's growing interest in Gallipoli: political leadership and a response to enthusiastic antipodean commemoration.[141] There is substantial interest in the victory at Gallipoli in Turkey. Its message of territorial invincibility that 'Çanakkale is impassable' has passed into popular culture and political rhetoric. Mehmetçik has been presented as the source of the country's military power and Atatürk as the leader of that power, particularly in periods when the cult of Atatürk was as its height (1920s and 1930s, 1980s). Gallipoli is a key part of Atatürk's backstory, but it is not the most important part—that is reserved for the War of Independence, and as such Gallipoli does not amount to a nation-building myth. As a moment of victory amidst the cumulative humiliations of the last days of the Ottoman Empire, it is perhaps too compromised by its context to be a perfect equivalent to Anzac Day in the southern hemisphere.

9

Conclusion

Australia and its Anzac legend dominate the memory and commemoration of the Gallipoli campaign. This book has argued that Anzac Day is Australia's greatest export: inspiring key aspects of New Zealand's arrangements, overshadowing Britain's own commemorations of Gallipoli, and surviving in Ireland at times when other memorials of the First World War were overgrown. All of Australia's subsequent involvement in war has been incorporated into its Anzac Day arrangements. Such is the importance of this first blooding in war for the nation that it overshadows all of its major engagements in the war: Pozières, Villers-Bretonneux, Beersheba, and Gaza. It has subsequently been exported to locations as diverse and distant from Gallipoli as 'Hell Fire Pass' (Thailand), Sandakan (Malaysia), and Long Tan (Vietnam). Thus Gallipoli dominates the Australian memory of war, even at the expense of the Western Front, which for Australia, as it was for New Zealand, was the far longer and more deadly part of the First World War.

Australia has also led the way in the revival of interest in the campaign. Whereas Australia's civil religion was born again in 1990 at Anzac Cove, New Zealand followed suit in the following decade. Ireland has been on its own specific path in rediscovering its role in the First World War, and Britain's burgeoning interest in the war has become increasingly apparent since around the turn of the century, but no distinctive or widespread interest in Gallipoli has emerged beyond very specific local and regimental connections in these two countries. Most significantly, Turkey is also participating in this

Gallipoli revival. For its own reasons and in response to the intense Australian and New Zealand interest in one small part of their territory, it is increasingly committed to commemorating the campaign.

The narrative of the campaign presented here has sought to acknowledge its multinational and multi-ethnic dimensions. That has meant not merely placing the Anzacs in what should be the familiar context of the Mediterranean Expeditionary Force (MEF) with its soldiers and sailors from across the British and French empires, but also portraying the MEF as a force that invaded the Ottoman Empire. The victorious Ottoman forces were themselves from a diverse range of ethnicities, and they owed their success not just to errors and failings by the MEF, but also to their own skilful and tenacious defence of their homeland. The manner in which that victory has been remembered in the Ottoman Empire and in Turkey has been presented here alongside other belligerent nations for the first time. It reveals the difficulties of remembering an isolated victory amidst a series of humiliating defeats, and the way in which the role of Mustafa Kemal has, at times, been elevated above all other aspects of the Ottoman campaign. Ultimately, although fellow feeling and Korean War-era gestures existed, it has been Kemal's extraordinary 1934 message that has proved to be the basis of the increasingly warm relations between Turkey and Australia and New Zealand since the 1980s. The depoliticizing focus on the human story at the heart of war has been replicated in Ireland's recovery of its First World War memory, and will surely feature in the careful diplomatic footwork that will be required in all of the centenary's international commemorations.

Yet if the military narrative attempted to reflect the campaign's diversity, it has not been possible to investigate all of the places and ways in which Gallipoli has been commemorated or remembered. The greatest absence is France, and this reflects the forgetfulness about the campaign there. A memorial service for the campaign was held in Paris on 25 April 1920, but it was a matter for L'Association des Anciens Combattants des Dardanelles, a veterans' group which also organized a pilgrimage to Gallipoli in 1930 and who met in 1935 to rekindle the torch at the Tomb of the Unknown Soldier at the Arc de Triomphe in Paris.[1] But what of France's colonial soldiers? How is Gallipoli remembered in Senegal? How has a postcolonial, multifaith memory unfolded in West Africa? It has been observed that in the French cemetery near S beach at Morto Bay, 'many of French colonial troops buried there shared the Islamic faith of their Turkish enemies. Each soldier was "remembered" beneath a memorial cross, the symbol of an alien religion.'[2] In January 2013, Prime Minister Erdoğan visited Senegal as part of Turkey's wider 'Africa initiative', and commented that the imperialist West had made the Turks and Senegalese fight against each other.[3] Did he strike a chord, or has Gallipoli been forgotten there as in metropolitan France? For the remainder of the MEF, too little is known about the Indian part in the campaign, or of the traces of memory among veterans and their families.[4] And it sometimes seems as if we know more about the Zion Mule Corps and its Irish commander, John Patterson,[5] or the Newfoundlanders,[6] than we do about the men from the subcontinent, or from Ceylon, Germany, or Austria-Hungary.

The transnational history of the campaign's memory that has been told here reflects on the politics of commemoration and the relationship between commemoration and national identity. At first the commemoration of Gallipoli reflected the inner workings of the British Empire—the shared cultural heritage which led commentators around the world to draw on the common vocabulary of Christian liturgy, classical references, and the purple prose of imperial rhetoric. To celebrate the sacrifices at Gallipoli of local boys was to burnish the empire's reputation. No distinction was drawn between national and imperial achievements. But the study of Gallipoli's commemoration thereafter reveals the unravelling of Britishness. In Ireland, this was rapidly and violently under way from 24 April 1916. In Australia, this occurred steadily and proudly throughout the interwar period, with the death knell for Britishness sounding from Singapore in 1942 onwards. In New Zealand, the break with Britain appears to have been more reluctant, hastened by the mother country's turn to Europe in 1973. That leaves Britain walking a post-imperial tightrope. Where it once submerged distinctive local English or Scottish links to the campaign in a bid to boost the Dominions, it will now find friendly foreign countries dominating its commemoration in London of the centenary of one of the major campaigns of the war.

From among the belligerents at Gallipoli, four new nations emerged. Two of them, Australia and New Zealand, built on the sacrifice in the campaign to develop a sense of themselves as independent nations. In Turkey, Gallipoli was of limited utility for this same purpose. In Ireland it was redundant, even threatening, to the new cause. The present-minded nature of commemoration is thereby demonstrated by Gallipoli in manifold ways. At first it was urgently required by grieving families, but it was also part of the continual mobilization for the war effort. Thereafter, in honouring the sacrifices of war, other meanings could be attached. In the Ottoman Empire, for example, these meanings could shift from a celebration of Mehmetçik's defence of the Sultanate to the elevation of the inspirational leadership of the Republic's founding president, Mustafa Kemal. After the Second World War, the campaign seemed less relevant, more problematic, and was an opportunity for criticism of the political situation of the day in those countries where the campaign's memory enjoyed a national profile. Since that low point, the Anzac revival in Australia has reflected the completeness of the break with Britain and the remarkable reorientation towards the former enemy. Surely, no other campaign of the First World War can claim to have been remembered in so many diverse fashions, to have captured the imagination to such an extent, and to have thereby moulded its belligerent nations' sense of themselves in such interesting and pervasive ways.

NOTES

Preface

1. Erik Zürcher, *The Young Turk Legacy and Nation Building: From the Ottoman Empire to Ataturk's Turkey* (London, 2010), 166–7.
2. Edward J. Erickson, *Ordered to Die: A History of the Ottoman Army in the First World War* (Westport, CT, 2000); Tim Travers, *Gallipoli 1915* (Stroud, 2001); Robin Prior, *Gallipoli: The End of the Myth* (New Haven, 2009); Edward J. Erickson, *Gallipoli: The Ottoman Campaign* (Barnsley, 2010). Mesut Uyar's work suggests that there are at least 30 Turkish war memoirs of the campaign, often written by career and reserve officers (personal communication from Mesut Uyar).
3. Jenny Macleod, *Reconsidering Gallipoli* (Manchester, 2004).
4. Please see the Newspaper section of the Bibliography for information on the main digitized historical newspapers and repositories of collected digitized historical newspapers that have been used.
5. Pierre Nora, 'General Introduction: Between Memory and History', in Nora (ed.), *Realms of Memory: Rethinking the French Past*, i. *Conflicts and Divisions* (New York, 1996), 3.
6. Roy Foster, 'Remembering 1798', in Ian McBride (ed.), *History and Memory in Modern Ireland* (Cambridge, 2001), 68.
7. One author who disapproves of using invasion/invaded is John A. Moses, 'Gallipoli or Other Peoples' Wars Revisited: Sundry Reflections on Anzac: A Review Article', *Australian Journal of Politics and History* 57/3 (2011), 434.

Chapter 1

1. The attack involved, from north to south, men from Mustafa Kemal's 19th Division, the 5th Division, fresh reinforcements from 2nd Division, 16th Division, and in the south the 77th (Arab) Regiment.
2. Edward J. Erickson, *Gallipoli: The Ottoman Campaign* (Barnsley, 2010), 102, 106–7.
3. Robin Prior, *Gallipoli: The End of the Myth* (New Haven, 2009), 128.
4. Otto Liman von Sanders, *Five Years in Turkey*, trans. US Army (Retired) Colonel Carl Reichmann (Annapolis, MD, 1927), 76.

5. Peter Stanley, *Quinn's Post: Anzac, Gallipoli* (Sydney, 2005), 66.
6. Henry W. Nevinson, *Last Changes, Last Chances* (London, 1928), 37.
7. Compton Mackenzie, *Gallipoli Memories* (London, 1929), 83.
8. Aubrey Herbert, *Mons, Anzac and Kut* (London, 1919), 115.
9. Mark Harrison, *The Medical War: British Military Medicine in the First World War* (Oxford, 2010), 178.
10. For details of the changing nature of the Simpson legend, see Peter Cochrane, *Simpson and the Donkey: The Making of a Legend* (Melbourne, 1992).
11. Judith Ireland, 'Aussies Bid at the Last Minute for Gallipoli Centenary', *Sydney Morning Herald*, 2 February 2014.
12. Warren Snowdon, MP, 'House of Representatives Ministerial Statements: Anzac Centenary', *Commonwealth of Australia Parliamentary Debates*, 16 May 2013.
13. One group of eminent Australian historians were moved to write about the dominance of Anzac Day in Australian national life: Marilyn Lake, Henry Reynolds, Mark McKenna, and Joy Damousi (eds), *What's Wrong with Anzac? The Militarisation of Australian History* (Sydney, 2010).
14. Uğur Ümit Üngör, *The Making of Modern Turkey: Nation and State in Eastern Anatolia, 1913–1950* (Oxford, 2011), 218–32.
15. Jenny Macleod, 'Britishness and Commemoration: National Memorials to the First World War in Britain and Ireland', *Journal of Contemporary History* 48/4 (2013), 663.
16. The author noted this pattern a decade ago; since then, interest has grown seemingly inexorably. Jenny Macleod, 'The Fall and Rise of Anzac Day: 1965 and 1990 Compared', *War & Society* 20/1 (May 2002), 149–68.
17. Paul Daley, 'Australia Spares No Expense as the Anzac Legend Nears Its Century', *The Guardian*, 14 October 2013, www.theguardian.com; 'Funding the Centenary Programme, WW100 New Zealand', www.ww100.govt.nz/about/about-WW100/funding-allocation#.UyXAHl5Bx74, accessed 16 March 2014.
18. John Horne, 'Our War, Our History', in Horne (ed.), *Our War: Ireland and the Great War. The 2008 Thomas Davis Lecture Series* (Dublin, 2008), 3–14.
19. 'Decade of Centenaries', www.decadeofcentenaries.com/, accessed 6 April 2014.
20. 'Çanakkale Zaferinin 50. Yıldönümü', *Milliyet*, 19 March 1965, p. 3; 'İrlandalı Muharipler Çanakkale'ye Geliyor', *Milliyet*, 23 March 1965, p. 4; 'Anzaklar Geldi', *Milliyet*, 22 April 1965, p. 1. The Anzac pilgrimage was described in detail by Ken Inglis who accompanied the trip: K. S. Inglis, 'Return to Gallipoli' in Inglis, *Anzac Remembered: Selected Writings of K. S. Inglis*, ed. John Lack (Melbourne, 2001).
21. The extensive literature on the Australian performance and experience at Gallipoli includes: Bill Gammage, *The Broken Years: Australian Soldiers in the Great War* (Ringwood, Victoria, 1974; repr. 1975); Dale Blair, *Dinkum Diggers:*

An Australian Battalion at War (Melbourne, 1997); Stanley, *Quinn's Post*; Rhys Crawley, *Climax at Gallipoli: The Failure of the August Offensive* (Norman, OK, 2013). The vast literature on the Australian aspects of the Anzac legend—that is, the memory of the campaign—includes: Kenneth Stanley Inglis, 'The Australians at Gallipoli, 2 Pts', *Historical Studies* 14 (1970), 219–30 and 361–75; Kevin Fewster, 'Ellis Ashmead-Bartlett and the Making of the Anzac Legend', *Journal of Australian Studies* 10 (1982), 17–30; Inglis, *Anzac Remembered*, ed. Lack; Alistair Thomson, *Anzac Memories: Living with the Legend* (Melbourne, 1994); Jenny Macleod, *Reconsidering Gallipoli* (2004); Bruce C. Scates, *Return to Gallipoli: Walking the Battlefields of the Great War* (Cambridge, 2006).

22. Sir Martin Gilbert, 'Introduction', in Gilbert, *The Straits of War: Gallipoli Remembered* (Stroud, 2000), p. xxi.

23. Ayhan Aktar, '18 Mart Zaferi'nin Unutulan Kahramanları Yahut, Çanakkale Bir "Haçlı Seferi" Midir', *Taraf*, 18 March 2014.

24. Hew Strachan, 'The First World War as a Global War', *First World War Studies* 1/1 (2010), 3–14.

Chapter 2

1. Aksakal gives the following list of losses since 1878: Cyprus; Ardahan, Batum, and Kars; Montenegro, Romania, and Serbia; Bosnia-Herzegovina; Tunisia; Egypt; Crete; Kuwait; Bulgaria; Tripoli; Dodecanese Islands; western Thrace; Aegean islands including Chios and Mitylene; Albania; Macedonia. Mustafa Aksakal, *The Ottoman Road to War in 1914: The Ottoman Empire and the First World War* (Cambridge, 2008), 5.

2. For an overview of Ottoman history (despite the title), see Norman Stone, *Turkey: A Short History* (London, 2010).

3. Aksakal, *The Ottoman Road to War*, 102, 153.

4. Chris B. Rooney, 'The International Significance of British Naval Missions to the Ottoman Empire, 1908–1914', *Middle Eastern Studies* 34/1 (1998), 1–29; Otto Liman von Sanders, *Five Years in Turkey*, trans. US Army (Retired) Colonel Carl Reichmann (Annapolis, MD, 1920; repr. 1927), 1.

5. Aksakal, *The Ottoman Road to War*, 91.

6. Winston S. Churchill to Sir Archibald Moore (Rear-Admiral, Third Sea Lord), 1 August 1914 in Martin Gilbert, *Winston S. Churchill*, iii. *Companion Part I, Documents July 1914–April 1915* (London, 1972), 9 (hereafter CV iii).

7. Henry Morgenthau, *Ambassador Morgenthau's Story: A Personal Account of the Armenian Genocide* (New York, 1918; repr. 2010), 48–57.

8. For the most incisive critique of decision-making in London regarding the Dardanelles, see Robin Prior, *Gallipoli: The End of the Myth* (New Haven, 2009).

9. Robert Rhodes James, *Gallipoli* (London, 1965; repr. 1999), 10–11.

10. Major General Callwell, Memorandum, 3 September 1914, in Gilbert, CV iii. 80–1.
11. Peter Hart, *Gallipoli* (London, 2011), 13.
12. Lord Kitchener to Winston S. Churchill, 2 January 1915, in Gilbert, CV iii. 360–1.
13. Lord Fisher to Winston S. Churchill, 3 January 1915, in Gilbert, CV iii. 367.
14. Rhodes James, *Gallipoli*, 27–8.
15. Vice Admiral Carden to Churchill, 5 January 1915, in Gilbert, CV iii. 380.
16. Carden to Churchill, 11 January 1915, in Gilbert, CV iii. 405.
17. Churchill to Carden, 14 January 1915, in Gilbert, CV iii. 415–16.
18. Churchill to Grand Duke Nicholas, 19 January 1915, in Gilbert, CV 3, p. 430.
19. Meeting of the War Council: Extract from Secretary's Notes, in Gilbert, CV iii. 463.
20. Keith Neilson, 'Kitchener, Horatio Herbert, Earl Kitchener of Khartoum (1850–1916)', in *Oxford Dictionary of National Biography* (Oxford, 2004), 41; George Cassar, 'Kitchener at the War Office', in Hugh Cecil and P. H. Liddle (eds), *Facing Armageddon: The First World War Experienced* (London, 1996).
21. Fisher to Sir John Jellicoe, 21 January 1915, in Gilbert, CV iii. 436.
22. Evidence of Admiral Lord Fisher of Kilverstone, Wednesday 11 October 1916 (day 10), Dardanelles Commission, q.3364, London, The National Archives, CAB 19/33.
23. Meeting of the War Council: Extract from Secretary's Notes, 13 January 1915, in Gilbert, CV iii. 411.
24. Edward J. Erickson, 'Strength Against Weakness: Ottoman Military Effectiveness at Gallipoli, 1915', *Journal of Military History* 65 (2001), 983. For brief details of this fleet and the ensuing coup, see Aksakal, *The Ottoman Road to War*, 78–9.
25. Tim Travers, *Gallipoli 1915* (Stroud, 2001), 21.
26. Lieutenant Colonel Hankey, Memorandum, 28 December 1914, in Gilbert, CV iii. 341–2.
27. Churchill to Fisher, 4 January 1915, in Gilbert, CV iii. 371.
28. George H. Cassar, *The French and the Dardanelles: A Study of Failure in the Conduct of War* (London, 1971), 49, 63.
29. Meeting of War Council: Conclusions, 16 February 1915, in Gilbert, CV iii. 516.
30. Meeting of the War Council: Secretary's Notes, 19 February 1915, in Gilbert, CV iii. 527–34.
31. Meeting of the War Council: Extract from Secretary's Notes, 24 February 1915, in Gilbert, CV iii. 557.
32. Prior, *Gallipoli*, 61–2.
33. Cassar, *The French and the Dardanelles*, 79, 60, 74.
34. Prior, *Gallipoli*, 66.

35. Meeting of the War Council: Secretary's Notes, 10 March 1915, in Gilbert, *CV* iii. 664.

36. Sean McMeekin, *The Berlin–Baghdad Express: The Ottoman Empire and Germany's Bid for World Power, 1898–1918* (London, 2010), 181.

37. Edward J. Erickson, *Gallipoli: The Ottoman Campaign* (Barnsley, 2010), 8–9.

38. Many of these were assigned to the Ottoman Navy's headquarters, so only about half of these men were assigned to the Dardanelles.

39. Mesut Uyar and Edward J. Erickson, *A Military History of the Ottomans: From Osman to Atatürk* (Santa Barbara, CA, 2009), 258.

40. Erickson, *Gallipoli*, 13; Erickson, *Ordered to Die: A History of the Ottoman Army in the First World War* (Westport, CT, 2000), 78–9.

41. Rudenno states that 'the *Majestic, Canopus, Formidable* and *Duncan* class battleships were all marked for decommissioning within fifteen months'. That would have meant HMS *Majestic*, HMS *Prince George*, HMS *Albion*, HMS *Canopus*, HMS *Goliath*, HMS *Ocean*, HMS *Vengeance*, HMS *Irresistible*, HMS *Implacable*, and HMS *Cornwallis*. Victor Rudenno, *Gallipoli: Attack from the Sea* (New Haven, 2008), 16.

42. Admiralty to Vice Admiral Carden, 'Dardanelles Operation Orders', 5 February 1915, in Gilbert, *CV* iii. 485. The submarines were the only vessels which did manage to pass through the straits. The first to do so was the *B11* on 13 December. Rudenno, *Gallipoli*, 19–23.

43. Prior, *Gallipoli*, 45–50.

44. Prior, *Gallipoli*, 53; Rudenno, *Gallipoli*, 39; Erickson, *Gallipoli*, 18–19.

45. McMeekin, *The Berlin–Baghdad Express*, 185.

46. Hans Kannengiesser Pasha, *The Campaign in Gallipoli*, trans. Major C. J. P. Ball DSO MC (Late RA) (London, 1928), 114.

47. Erickson, *Gallipoli*, 18.

48. R. Swing, *Good Evening! Raymond 'Gram' Swing* (1965), 71, quoted in Rudenno, *Gallipoli*, 50.

49. Hart, *Gallipoli*, 37–9; Rudenno, *Gallipoli*, 51.

50. Travers, *Gallipoli 1915*, 31.

51. Prior, *Gallipoli*, 57.

52. Erickson, *Gallipoli*, 20–1.

53. Hart, *Gallipoli*, 33.

54. Erickson, *Gallipoli*, 26. See also Travers, *Gallipoli 1915*, 31–2.

55. Erickson, *Gallipoli*, 29.

Chapter 3

1. Unpublished lectures: Nigel Steel, '"Heroic Sacrifice": The Sikh Regiment at Gallipoli, June 1915', Imperial War Museum North, 2004; Peter Stanley, 'Australia, India and Gallipoli: A Study in Contrasting National Consequences', Australian War Memorial, 2004.

2. Michael Hickey, *Gallipoli* (London, 1995), 59.

3. Peter Hart, *Gallipoli* (London, 2011), 47–51.

4. George H. Cassar, *The French and the Dardanelles: A Study of Failure in the Conduct of War* (London, 1971), 88.

5. Mesut Uyar, 'Ottoman Arab Officers between Nationalism and Loyalty During the First World War', *War in History* 20/4 (2013), 537.

6. Salim Tamari, *Year of the Locust: A Soldier's Diary and the Erasure of Palestine's Ottoman Past* (Berkeley and Los Angeles, 2011), 12–13; Uyar, 'Ottoman Arab Officers', 539.

7. Mesut Uyar and Ed Erickson, *A Military History of the Ottomans: From Osman to Atatürk* (Santa Barbara, CA, 2009), 259.

8. Sean McMeekin, *The Berlin–Baghdad Express: The Ottoman Empire and Germany's Bid for World Power, 1898–1918* (London, 2010), 183; Uyar and Erickson, *A Military History of the Ottomans*, 238.

9. Erickson, *Gallipoli: The Ottoman Campaign* (Barnsley, 2010), 36–8.

10. John Lee, *A Soldier's Life: General Sir Ian Hamilton 1853–1947* (London, 2000), 144–7.

11. Cassar, *The French and the Dardanelles*, 78, 121.

12. Erickson, *Gallipoli*, 7–16.

13. Erickson, *Gallipoli*, 30–5.

14. Otto Liman von Sanders, 'The Campaign of Gallipoli, 1915–1916', *Royal United Services Institution Journal* 67/465 (1922), 147.

15. Erickson, *Gallipoli*, 39. There were 22,000 Greeks living on the peninsula.

16. Erickson, *Ordered to Die: A History of the Ottoman Army in the First World War* (Westport, CT, 2000), 81–2.

17. Tim Travers, 'Liman Von Sanders, the Capture of Lieutenant Palmer, and Ottoman Anticipation of the Allied Landings at Gallipoli on 25 April 1915', *Journal of Military History* 65 (2001), 979.

18. Uyar and Erickson, *A Military History of the Ottomans*, 258.

19. Liman von Sanders, *Five Years in Turkey*, trans. US Army (Retired) Colonel Carl Reichmann (Annapolis, MD, 1927), 63.

20. Sir I. S. M. Hamilton, *Gallipoli Diary*, i (London, 1920), 14.

21. P. Chasseaud and P. Doyle, *Grasping Gallipoli: Terrain, Maps and Failure at the Dardanelles, 1915* (Staplehurst, 2005).

22. Tim Travers, *Gallipoli 1915* (Stroud, 2001), 66.

23. Travers, *Gallipoli 1915*, 68–9; Erickson, *Gallipoli*, 49.

24. Robin Prior, *Gallipoli: The End of the Myth* (New Haven, 2009), 111–12.

25. Diary of Lieutenant Colonel S. P. Weir, cited in Harvey Broadbent, *Gallipoli: The Fatal Shore* (Camberwell, 2005), 56.

26. Prior, *Gallipoli*, 115–16.

27. Corporal Adil, interview with Harvey Broadbent, 1985, cited in Broadbent, *Gallipoli*, 55.

28. Erickson, *Gallipoli*, 50.

29. Travers, *Gallipoli 1915*, 70–1.
30. Prior, *Gallipoli*, 115.
31. Travers, *Gallipoli 1915*, 71.
32. Prior, *Gallipoli*, 117.
33. Erickson, *Gallipoli*, 52–5. During this time, Esat Pasha, commander of III Corps, reorganized his commanders' responsibilities to reflect how matters had developed. He freed up Halil Sami to focus on Cape Helles, by detaching Sefik's 27th Regiment from his forces and passing them to Kemal, who was now designated as Ari Burnu Front commander.
34. Prior, *Gallipoli*, 118.
35. Erickson, *Ordered to Die*, 83.
36. Erickson, *Gallipoli*, 54.
37. Travers, *Gallipoli 1915*, 76.
38. Erickson, *Gallipoli*, 56.
39. Hamilton, *Gallipoli Diary*, i. 144.
40. Tim Travers, *The Killing Ground: The British Army, the Western Front and the Emergence of Modern Warfare 1900–1918* (Barnsley, 1987; repr. 2003), 45.
41. Erickson, *Gallipoli*, 56; Prior, *Gallipoli*, 119.
42. Travers, *Gallipoli 1915*, 49–50, 58–61.
43. Erickson, *Gallipoli*, 65–9.
44. Erickson, *Gallipoli*, 69–71.
45. Hart, *Gallipoli*, 144.
46. Captain Guy Geddes, report, Royal Munster Fusiliers War Diary, cited in Hart, *Gallipoli*, 149.
47. Travers, *Gallipoli 1915*, 54.
48. Erickson, *Gallipoli*, 72.
49. Travers, *Gallipoli 1915*, 75.
50. Erickson, *Ordered to Die*, 84; Erickson, *Gallipoli*, 73–5.
51. Uyar and Erickson, *A Military History of the Ottomans*, 260.
52. Hart, *Gallipoli*, 164–7.
53. Hart, *Gallipoli*, 176–7.
54. Travers, 'Liman Von Sanders', 979.
55. Erickson, *Gallipoli*, 82–6; Hart, *Gallipoli*, 170–5.
56. Victor Rudenno, *Gallipoli: Attack from the Sea* (New Haven, 2008), 70.
57. Erickson, *Gallipoli*, 56.
58. Prior, *Gallipoli*, 108.
59. Edward J. Erickson, 'Strength Against Weakness: Ottoman Military Effectiveness at Gallipoli, 1915', *Journal of Military History* 65 (2001), 1007–8.
60. Prior, *Gallipoli*, 108.
61. Les Carlyon, *Gallipoli* (Sydney, 2001), 232.
62. Nigel Steel and Peter Hart, *Defeat at Gallipoli* (London, 2002), 137.
63. Erickson, *Gallipoli*, 67.
64. Travers, *Gallipoli 1915*, 56–8.

65. Hart, *Gallipoli*, 102–3; Robert Rhodes James, *Gallipoli* (London, 1965; repr. 1999), 110.
66. Uyar and Erickson, *A Military History of the Ottomans*, 241.
67. Erik J. Zürcher, *The Young Turk Legacy and Nation Building: From the Ottoman Empire to Ataturk's Turkey* (London, 2010), 170; Erickson, *Gallipoli*, 40.
68. Erickson, 'Strength Against Weakness', 1003.
69. Travers, *Gallipoli 1915*, 61.
70. Erickson, *Ordered to Die*, 85.

Chapter 4

1. Robin Prior, *Gallipoli: The End of the Myth* (New Haven, 2009), 131–3; Tim Travers, *Gallipoli 1915* (Stroud, 2001), 93.
2. Edward J. Erickson, *Gallipoli: The Ottoman Campaign* (Barnsley, 2010), 81.
3. Prior, *Gallipoli*, 133.
4. Erickson, *Gallipoli*, 94–7.
5. Mesut Uyar and Edward J. Erickson, *A Military History of the Ottomans* (Santa Barbara, CA, 2009), 260.
6. Prior, *Gallipoli*, 140–4; Travers, *Gallipoli 1915*, 102–3.
7. Erickson, *Gallipoli*, 100.
8. The attack is described in Chapter 1, 'Introduction'.
9. Travers, *Gallipoli 1915*, 105.
10. Prior, *Gallipoli*, 151.
11. Travers, *Gallipoli 1915*, 108.
12. Edward J. Erickson, *Ordered to Die: A History of the Ottoman Army in the First World War* (Westport, CT, 2000), 87–8; Erickson, *Gallipoli*, 110.
13. Prior, *Gallipoli*, 154.
14. Erickson, *Gallipoli*, 125, 128.
15. Peter Hart, *Gallipoli* (London, 2011), 231–4.
16. Mark Harrison, *The Medical War: British Military Medicine in the First World War* (Oxford, 2010), 172–82.
17. Andrea McKenzie, '"Our Common Colonial Voices": Canadian Nurses, Patient Relations, and Nation on Lemnos', in Joachim Bürgschwentner et al. (eds), *Other Fronts, Other Wars? First World War Studies on the Eve of the Centennial* (Leiden, 2014). See also Janet Butler, 'Nursing Gallipoli: Identity and the Challenge of Experience', *Journal of Australian Studies* 27/78 (2003), 47–57.
18. Harrison, *The Medical War*, 195.
19. Erickson, *Gallipoli*, 195.
20. Erickson, *Gallipoli*, 191; Erik J. Zürcher, *The Young Turk Legacy and Nation Building: From the Ottoman Empire to Ataturk's Turkey* (London, 2010), 180.
21. Erickson, *Gallipoli*, 197.
22. Erickson, *Gallipoli*, 136–7.

23. Rhys Crawley, *Climax at Gallipoli: The Failure of the August Offensive* (Norman, OK, 2013), 189; Hart, *Gallipoli*, 286–91.
24. Erickson, *Gallipoli*, 145–7; Travers, *Gallipoli 1915*, 117; Crawley, *Climax at Gallipoli*, 191.
25. Travers, *Gallipoli 1915*, 114.
26. Erickson, *Gallipoli*, 150–2; Prior, *Gallipoli*, 174.
27. Bill Gammage, *The Broken Years: Australian Soldiers in the Great War* (Ringwood, Victoria, 1974; repr. 1975), 74.
28. Prior, *Gallipoli*, 267 n. 18; Travers, *Gallipoli 1915*, 118.
29. Prior, *Gallipoli*, 178–9.
30. Erickson, *Gallipoli*, 152–3.
31. Erickson, *Gallipoli*, 164–6.
32. Glyn Harper (ed.), *Letters from Gallipoli: New Zealand Soldiers Write Home* (Auckland, 2011), 240–1.
33. Hart, *Gallipoli*, 321–5; Sean McMeekin, *The Berlin–Baghdad Express: The Ottoman Empire and Germany's Bid for World Power, 1898–1918* (London, 2010), 187.
34. Travers, *Gallipoli 1915*, 138–41, 145; Hart, *Gallipoli*, 280–1.
35. Prior, *Gallipoli*, 196.
36. Prior, *Gallipoli*, 192.
37. Hart, *Gallipoli*, 350–3; Crawley, *Climax at Gallipoli*, 203–4.
38. Erickson, *Gallipoli*, 158; Erickson, *Ordered to Die*, 90.
39. Erickson, *Gallipoli*, 139.
40. Prior, *Gallipoli*, 206; Travers, *Gallipoli 1915*, 150.
41. Crawley, *Climax at Gallipoli*, 206–7.
42. Erickson, *Gallipoli*, 167; Travers, *Gallipoli 1915*, 160.
43. Travers, *Gallipoli 1915*, 162; Prior, *Gallipoli*, 207.
44. Crawley, *Climax at Gallipoli*, 213.
45. Erickson, *Gallipoli*, 166.
46. Crawley, *Climax at Gallipoli*, 214–41.
47. Crawley, *Climax at Gallipoli*, 56, 125.
48. Erickson, *Gallipoli*, 182.
49. Travers, *Gallipoli 1915*, 203–5.
50. Jenny Macleod, 'Ellis Ashmead-Bartlett, War Correspondence and the First World War', in Y. T. McEwen and F. A. Fisken (eds), *War, Journalism and History: War Correspondents in the Two World Wars* (Bern, 2012), 62.
51. Erickson, *Gallipoli*, 176–80.
52. Erickson, *Gallipoli*, 180.
53. Erickson, *Ordered to Die*, 94.
54. Gürsel Göncü and Şahin Aldoğan, *Siperin Ardı Vatan* (Istanbul, 2006), 146.
55. Edward J. Erickson, 'Strength Against Weakness: Ottoman Military Effectiveness at Gallipoli, 1915', *Journal of Military History* 65 (2001), 1010.

56. Göncü and Aldoğan, *Siperin Ardı Vatan*, 146. Mesut Uyar suggests a much lower figure of 166,507 casualties overall. Uyar and Erickson, *A Military History of the Ottomans*, 261.
57. Prior, *Gallipoli*, 242.
58. Crawley, *Climax at Gallipoli*, 59.
59. Erickson, *Ordered to Die*, 89–90.
60. Erickson, 'Strength Against Weakness', 1010.
61. Uyar and Erickson, *A Military History of the Ottomans*, 259.

Chapter 5

1. A Special Correspondent [Ellis Ashmead-Bartlett], 'The Gallipoli Landing', *The Times*, 7 May 1915, p. 7.
2. 'Attack on the Straits', *The Times*, 27 April 1915, p. 8.
3. S. A. Moseley, *The Truth About a Journalist* (London, 1935), 101.
4. K. Fewster, 'Ellis Ashmead-Bartlett and the Making of the Anzac Legend', *Journal of Australian Studies*, 10 (June 1982), 30.
5. C. E. W. Bean, *The Story of Anzac from the Outbreak of War to the End of the First Phase of the Gallipoli Campaign, May 4, 1915*, i (St Lucia, Queensland, 1921; repr. 1981); C. E. W. Bean, *The Story of Anzac: From 4 May, 1915 to the Evacuation of the Gallipoli Peninsula*, ii (St Lucia, Queensland, 1924; repr. 1981).
6. C. E. W. Bean, 'The Writing of the Australian Official History of the Great War: Sources, Methods and Some Conclusions', *Royal Australian Historical Society, Journal and Proceedings*, 24/2 (1938), 89.
7. E. M. Andrews, 'Bean and Bullecourt: Weaknesses and Strengths of the Official History of Australia in the First World War', *Revue internationale d'histoire militaire* 72 (August 1990), 43–5.
8. E. K. Bowden, 'Australian War Histories: Renewal of Engagement of Official Historian (Mr. C. E. W. Bean)', 4 June 1924, Bean Papers, AWM 38 3 DRL 6673, item 11.
9. Bean, 'The Writing of the Australian Official History of the Great War', 91.
10. Bean, *The Story of Anzac*, i. 4–5.
11. John Masefield, *Gallipoli* (London, 1916), 3.
12. Masefield, *Gallipoli*, 19. For an extended discussion of Masefield's book see Jenny Macleod, 'The British Heroic-Romantic Myth of Gallipoli', in Macleod (ed.), *Gallipoli: Making History* (London, 2004), 73–85.
13. *Irish Times*, 28 October 1916; *Sydney Morning Herald*, 11 November 1916.
14. See e.g. *Hawera & Normanby Star*, 24 April 1917; *Evening Post*, 24 April 1923; and *Auckland Star*, 23 April 1938.
15. Jenny Macleod, 'General Sir Ian Hamilton and the Dardanelles Commission', *War in History* 8/4 (2001), 418–41.
16. 'The Dardanelles Report', *The Times*, 19 November 1919.

17. D. A. Kent, 'The Anzac Book and the Anzac Legend: C. E. W. Bean as Editor and Image-Maker', *Historical Studies* 21/84 (April 1985), 390.

18. Joan Beaumont, 'The Anzac Legend', in Beaumont (ed.), *Australia's War 1914–1918* (St Leonards, NSW, 1995), 156.

19. E. Ashmead-Bartlett, *Ashmead-Bartlett's Despatches from the Dardanelles: An Epic of Heroism* (London, 1916). He also published *Some of My Experiences in the Great War* (London, 1918) and *the Uncensored Dardanelles* (London, 1928).

20. Ashmead-Bartlett Papers, W. H. Ifould, Principal Librarian, Public Library of New South Wales to Messrs. Angus and Robertson Ltd, 5 April 1916: ML A1583: 6. 'Ashmead-Bartlett in Sydney', *Sunday Times (Perth, WA)*, 13 February 1916.

21. 'Films of the War', *The Times*, 18 January 1916, p. 5; Philip Dutton, '"More Vivid Than the Written Word": Ellis Ashmead-Bartlett's Film, *With the Dardanelles Expedition* (1915)', *Historical Journal of Film, Radio and Television* 24/2 (2004), 205–22.

22. K. S. Inglis, 'Anzac, the Substitute Religion', in Craig Wilcox (ed.), *Observing Australia 1959 to 1999* (Carlton South, Victoria, 1999); Inglis, *Sacred Places: War Memorials in the Australian Landscape* (Carlton South, Victoria, 1998; repr. 2001), 458–63.

23. Commonwealth Bureau of Census and Statistics, *Year Book of the Commonwealth of Australia*, 6, (Melbourne, 1913), 103, 108.

24. Peter Hart, *Gallipoli* (London, 2011), 48.

25. Stuart Macintyre, *The Oxford History of Australia*, iv. *The Succeeding Age 1901–1942* (Melbourne, 1986; repr. 2001), 126.

26. Stephen Alomes, *A Nation at Last? The Changing Character of Australian Nationalism 1880–1988* (North Ryde, NSW, 1988), 41, 52.

27. Jay Winter, *The Great War and the British People* (London, 1986), 75.

28. Robin Prior, *Gallipoli: The End of the Myth* (New Haven, 2009), 242.

29. 'Australian Commonwealth Horse', in Peter Dennis et al. (eds), *The Oxford Companion to Australian Military History* (Melbourne, 1995), 65.

30. 'Mr. Ashmead-Bartlett's Story', *Sydney Morning Herald*, 8 May 1915.

31. John A. Moses, 'The Struggle for Anzac Day 1916–1930 and the Role of the Brisbane Anzac Day Commemoration Committee', *Journal of the Royal Australian Historical Society* 88/1 (2002), 55.

32. Inglis, *Sacred Places*, 75–80.

33. 'Anzac Day', *The Register*, 27 August 1915.

34. 'Anzac Day', *Sydney Morning Herald*, 14 October 1915.

35. 'Anzac Day', *Brisbane Courier*, 8 January 1916.

36. For example, he wrote to Perth City Council, 'Anzac Day', *West Australian*, 14 March 1916.

37. John Connor, *Anzac and Empire: George Foster Pearce and the Foundations of Australian Defence* (Cambridge, 2011), 62.

38. A cross-section of locations holding Anzac Day commemorations in 1916 were New South Wales: Bathurst, Dubbo, Moama, Moree, Sydney, Tamworth; Queensland: Brisbane, Ipswich, Rockhampton, Townsville; South Australia: Adelaide, Port Pirie, Mount Gambier; Tasmania: Hobart; Victoria: Bendigo, Camperdown, and Melbourne; Western Australia: Kalgoorlie and Perth. I have not found any references to events in Northern Territory.

39. 'For Australian Soldiers', *Referee*, 26 April 1916.

40. 'Parade of Troops', *Sydney Morning Herald*, 26 April 1916.

41. 'Anzac Day in Sydney', *Warwick Examiner and Times*, 26 April 1916.

42. 'Anzac Day', *Kadina and Wallaroo Times*, 26 April 1916.

43. 'Anzac Day', *The Mercury*, 13 April 1916.

44. 'Schools Commemorate Anzac Day', *Camperdown Chronicle*, 25 April 1916.

45. 'Anzac Day', *Riverine Herald*, 26 April 1916.

46. Moses, 'The Struggle for Anzac Day', 55. See also J. M. Winter, *Sites of Memory, Sites of Mourning: The Great War in European Cultural History* (Cambridge, 1995) which develops this argument to apply to the memory of the entire war.

47. 'Anzac Day', *The Argus*, 7 March 1916. Roman Catholic priests were forbidden from performing a Requiem Mass on 25 April until 1923. Inglis, 'Anzac, the Substitute Religion', 67.

48. 'Anzac Day', *Brisbane Courier*, 26 April 1916; 'To Get Men', *Sydney Morning Herald*, 4 April 1916. Over £5,000 was collected in Sydney: 'Over £5000 Collected', *Sydney Morning Herald*, 27 April 1916.

49. Martin Crotty and Craig Melrose, 'Anzac Day, Brisbane, Australia: Triumphalism, Mourning and Politics in Interwar Commemoration', *Round Table* 96/393 (2007), 685, 681.

50. 'Anzac Day', *Brisbane Courier*, 7 April 1916.

51. 'Workers and Anzac Day', *The Argus*, 19 April 1921.

52. Moses, 'The Struggle for Anzac Day', 71.

53. 'Anzac Day', *The Mercury*, 29 June 1920.

54. 'Anzac Sports at Cairo', *Sydney Morning Herald*, 3 June 1916.

55. 'Anzac Day', *The Examiner*, 14 April 1917.

56. 'To the Editor of the Herald', *Sydney Morning Herald*, 30 March 1916.

57. Joy Damousi, 'Private Loss, Public Mourning: Motherhood, Memory and Grief in Australia During the Inter-War Years', *Women's History Review* 8/2 (1999), 366.

58. Crotty and Melrose, 'Anzac Day, Brisbane', 684.

59. Damousi, 'Private Loss', 372.

60. It was renamed the Returned Services League of Australia (RSL) in 1965 to reflect what had become common usage.

61. Martin Crotty, 'The Anzac Citizen: Towards a History of the RSL', *Australian Journal of Politics and History* 53/2 (2007), 192.

62. Adrian Gregory, *The Silence of Memory* (Oxford, 1994), 40–1.

63. Inglis, *Sacred Places*, 241.
64. Moses, 'The Struggle for Anzac Day', 68.
65. Inglis, *Sacred Places*, 199.
66. 'The Celebration of Anzac', *Brisbane Courier*, 8 March 1916.
67. E. M. Andrews, *The Anzac Illusion: Anglo-Australian Relations During World War One* (Cambridge, 1993), 53–5.
68. Inglis, *Sacred Places* is an unparalleled survey of war memorials in Australia.
69. 'Anzac Day March of Troops to Cenotaph', *The Mercury*, 27 April 1925.
70. 'Anzac Day', *The Mercury*, 16 March 1925.
71. 'Anzac Day Speeches', *The Argus*, 18 March 1925.
72. 'Anzac Day', *Sydney Morning Herald*, 26 March 1925.
73. Inglis, *Sacred Places*, 207; Crotty and Melrose, 'Anzac Day, Brisbane', 684.
74. 'Anzac Day', *Sydney Morning Herald*, 25 April 1925.
75. Inglis, *Sacred Places*, 281.
76. Inglis, *Sacred Places*, 283–4.
77. Inglis, *Sacred Places*, 303–29.
78. 'Anzac Day: Rockhampton Commemoration', *Morning Bulletin*, 26 April 1916.
79. Australian Associated Press (hereafter AAP), 'Anzac Day Centenary: 2015 Year of Turkey in Australia', *Australian Times*, 27 April 2012.
80. Graham Seal, '"…And in the Morning…": Adapting and Adopting the Dawn Service', *Journal of Australian Studies* 35/1 (2011), 51–2.
81. 'At the Cenotaph', *Brisbane Courier*, 26 April 1929.
82. 'Anzac Day', *Sydney Morning Herald*, 8 April 1935.
83. 'Anzac Day', *Sydney Morning Herald*, 29 March 1935.
84. 'Anzac Day', *West Australian*, 8 April 1935.
85. 'For Soldiers Only Dawn Ceremony at Shrine', *The Argus*, 27 March 1935.
86. 'Anzac Day', *West Australian*, 25 April 1935.
87. 'Anzac Landing Commemoration Stamps', *The Argus*, 2 February 1935; Inglis, *Sacred Places*, 333. The national memorial in Australia had not been completed at this point.
88. 'These April Days!', *Sydney Morning Herald*, 20 April 1935.
89. 'Anzac rally 12,000 up on 1964', *The Sun*, Press cuttings: Anzac Day 1965–6 (Victoria), Australian War Memorial, Canberra (hereafter, Press cuttings, AWM).
90. Jane Ross, *The Myth of the Digger: The Australian Soldier in Two World Wars* (Sydney, 1985).
91. Deborah Gare, 'Britishness in Recent Australian Historiography', *Historical Journal* 43/4 (2000), 1151–2.
92. 'Anzac Split: It Must Not Happen Again', *Sydney Morning Herald*, 29 April 1956.
93. 'Anzac Day Service: Church Leaders Upset', *Mercury (Tas)*, 8 April 1965. Press cuttings, AWM.

94. 'Anzac Row Flares Up', *News*, 7 April 1965. Press cuttings, AWM.

95. 'Anzac Night in City Orderly, Say Police', *Sydney Morning Herald*, 26 April 1955.

96. 'Attack on Anzac Day Condemned', *Sydney Morning Herald*, 23 April 1960. For further details of post-war criticisms of Anzac Day see Carina Donaldson and Marilyn Lake, 'Whatever Happened to the Anti-War Movement?', in Lake et al. (eds), *What's Wrong with Anzac?* (Sydney, 2010), 79–93.

97. Damien Cash, 'Seymour, Alan', in William H. Wilde, Joy Hooton, and Barry Andrews (eds), *The Oxford Companion to Australian Literature* (Oxford: 1994).

98. 'Anzac Is for Peace', *News*, 26 April 1965.

99. Dr Barry Smith, 'Gallipoli: 50 Years After', *Sydney Morning Herald*, 24 April 1965.

100. 'Anzac, Fifty Years After', *The Sun-Herald*, 25 April 1965.

101. 'A Nation Pays Its Homage', *Sydney Morning Herald*, 26 April 1965.

102. Ken Inglis, 'Men, Women, and War Memorials: Anzac Australia', in Inglis, *Anzac Remembered: Selected Writings of K. S. Inglis*, ed. John Lack (Melbourne, 2001).

103. '340 Marchers Honour Women Raped in War', *Canberra Times*, 26 April 1983.

104. 'Minorities Attacked at Dawn Service', *Canberra Times*, 26 April 1988.

105. Dominic Bryan and Stuart Ward, 'The "Deficit of Remembrance": The Great War Revival in Australia and Ireland', in Katie Holmes and Stuart Ward (eds), *Exhuming Passions: The Pressure of the Past in Ireland and Australia* (Dublin, 2011), 163–86.

106. 'Commemoration of Anzac Day', *Canberra Times*, 26 April 1944.

107. 'Anzac Service Next Year in War Memorial Grounds', *Canberra Times*, 2 May 1945.

108. See e.g. 'Vast Crowd Sees Anzac Ceremony at War Memorial', *Canberra Times*, 26 April 1951. The 'vast crowd' in this instance was judged to be 3,000 strong. 'Record Crowd of 6,000 at Memorial Service', *Canberra Times*, 26 April 1952.

109. 'Dawn Service at War Memorial', *Canberra Times*, 26 April 1957.

110. 'Record Attendance at Anzac Commemoration', *Canberra Times*, 26 April 1957.

111. See e.g. 'Anzac Day', *Canberra Times*, 26 April 1969. The figures this year were 1,500 and 11,000 and the crowds were described as the biggest since 1965.

112. 'Anzac Day Old and Young, They All Came to Remember', *Canberra Times*, 26 April 1976.

113. 'Anzac Day 1984', *Canberra Times*, 26 April 1984.

114. *Australian War Memorial Annual Report 1991–1992* (1992).

115. 'Large crowds honour war veterans', *Canberra Times*, 26 April 1994.

116. '15,000 gather to honour, remember', *Canberra Times*, 26 April 1995.

117. *Australian War Memorial Annual Report 2005–2006* (2006), 22.

118. *Australian War Memorial Annual Report 2007–2008* (2008), 6.

119. Bill Gammage, *The Broken Years: Australian Soldiers in the Great War* (Ringwood, Victoria, 1974; repr. 1975); Patsy Adam-Smith, *The Anzacs* (London, 1978); Peter Weir, *Gallipoli* (Paramount, 1981).

120. Stuart Ward, '"A War Memorial in Celluloid": The Gallipoli Legend in Australian Cinema, 1940s–1980s,' in Jenny Macleod (ed.), *Gallipoli: Making History* (London, 2004), 64.

121. Bryan and Ward, '"Deficit of Remembrance"', 168–9.

122. Kent, '*The Anzac Book* and the Anzac Legend', 386.

123. 'Turkish Officer to Be Here on Anzac Day', *Canberra Times (ACT)*, 13 March 1953.

124. The first reference in a digitized newspaper that I have found to the speech is 'A Priceless Inheritance from Gallipoli', *Canberra Times*, 1 August 1981. The database has limited coverage of newspapers after 1955.

125. Brad West, 'Dialogical Memorialization, International Travel and the Public Sphere: A Cultural Sociology of Commemoration and Tourism at the First World War Gallipoli Battlefields', *Tourist Studies* 10/3 (2010), 214.

126. Catherine Simpson, 'From Ruthless Foe to National Friend: Turkey, Gallipoli and Australian Nationalism', *Media International Australia* 137 (2010), 63. Simpson gives the speech the elevated status of an 'ode'.

127. 'The Sacred Ground of Gallipoli', *Brisbane Courier*, 30 January 1923. See also Bart Ziino, *A Distant Grief: Australians, War Graves and the Great War* (Crawley, WA, 2007), 59–81.

128. George Davis, 'Turkey's Engagement with Anzac Day, 1948–2000', *War & Society* 28/2 (2009), 141.

129. Bart Ziino, 'Who Owns Gallipoli? Australia's Gallipoli Anxieties 1915–2005', *Journal of Australian Studies* 30/88 (2006), 1–12; Minister of Veterans Affairs Bruce Billson, 'Turkey Respects Battlefield', *Sunday Canberra Times*, 8 July 2007.

130. Text of Prime Minister's speech, Anzac Cove 25 April 1990, Press cuttings: Anzac Day 1990, Folder 62, AWM.

131. Nicholas Bromfield, 'Welcome Home: Reconciliation, Vietnam Veterans, and the Reconstruction of Anzac under the Hawke Government', www.auspsa.org.au/sites/default/files/welcome_home_nicholas_bromfield.pdf.

132. Mark McKenna, 'Anzac Day: How Did It Become Australia's National Day?', in Lake et al. (eds), *What's Wrong with Anzac?*, 113–16.

133. Jenny Macleod, 'The Fall and Rise of Anzac Day: 1965 and 1990 Compared', *War & Society* 20/1 (May 2002), 154.

134. James Curran, *The Power of Speech: Australian Prime Ministers Defining the National Image* (Carlton, Victoria, 2004), ch. 5.

135. James Curran and Stuart Ward, *The Unknown Nation: Australia after Empire* (Carlton, Victoria, 2010), 1–5, 241.
136. McKenna, 'Anzac Day', 123–5.
137. Daniel Hoare, 'Turks Allowed to Join Anzac March', The World Today (ABC Local Radio), 12 April 2006, http://www.abc.net.au/worldtoday/content/2006/s1614594.htm.
138. Tony Stephens, 'Time Marches Past', *Sydney Morning Herald*, 26 April 2004. Ex-Allied Forces veterans had joined Adelaide's parade by 1955, see J. G. Pavils, *Anzac Day: The Undying Debt* (Adelaide, 2007), 81.
139. Events for your Diary, http://indymedia.org.au/2012/04/24/canberra-syd ney-nsw-aboriginal-rights-events-for-your-diary-15-events-from-24-april-2012, accessed 22 May 2014.
140. David Huggonson, 'Equal in War, but Slow to Be Honoured by "White" Australians They Served', *Daily Advertiser*, 21 April 1990. News clippings: Anzac Day 1990, folder 11, AWM.
141. Estimate by Henry Schydlo, cameraman for ABC's live coverage in 1990 (personal communication).
142. Brad West, 'Enchanting Pasts: The Role of International Civil Religious Pilgrimage in Reimagining National Collective Memory', *Sociological Theory* 26/3 (2008), 261.
143. Errol Simper, 'ABC Takes the Nation to Gallipoli', *The Australian*, 11 April 1990. Further information from Henry Schydlo. The Dawn Service, the Lone Pine Ceremony, and the Chunuk Bair services were all broadcast.
144. 'The Governor-General's Anzac Day Speech', *Sydney Morning Herald*, 25 April 2004.
145. Macgregor Duncan et al., 'Youth's Image in Anzac Lore', *The Australian*, 25 April 2005.
146. John Howard, 'Anzac Day' (1996), www.pmtranscripts.dpmc.gov.au; 'Transcript of the Prime Minister the Hon John Howard MP, Address at Anzac Day Dawn Service, Gallipoli' (2005), www.pmtranscripts.dpmc. gov.au.
147. Julia Gillard, 'Dawn Service, Gallipoli' (2012), www.pmtranscripts.dpmc. gov.au.
148. Doug Conway, 'Battle on the Beach', *Canberra Times*, 24 December 2005; Lindy Edwards, 'New Era in Search for True Identity', *Canberra Times*, 12 February 2008.
149. '"A Place Shining with Honour"', *Sydney Morning Herald*, 25 April 2012.
150. Phillip Coorey, 'Day Embodies the Nation's Values, Says Gillard', *Sydney Morning Herald*, 26 April 2012.
151. AAP, 'Anzac Day Centenary: 2015 Year of Turkey in Australia', *Australian Times*, 27 April 2012.
152. 'Press Release: Anzac Centenary Program to Commemorate Australia's Military History' (2012), www.pmtranscripts.dpmc.gov.au.

153. James Brown, *Anzac's Long Shadow: The Cost of Our National Obseesion* (Collingwood, Victoria, 2014), 20, 4.
154. www.campgallipoli.com.au.
155. www.anzacrun.com.
156. Grace Koelma, 'RSL Fundraising Campaign Allows Australians to Pay to Listen to a Minute of Silence on Anzac Day', 13 April 2014. These examples of the commercialization of Anzac are drawn from the historypunk blog (www.historypunk.com) run by Jo Hawkins.
157. Marilyn Lake, 'Introduction', in Lake et al. (eds), *What's Wrong with Anzac?*, 4.

Chapter 6

1. 'Hoed the Turnips', *Auckland Star*, 8 June 1927.
2. Ian McGibbon, 'Anzac', in McGibbon (ed.), *The Oxford Companion to New Zealand Military History* (Auckland, 2000) (hereafter OCNZMH), 27.
3. Angela McCarthy, *Scottishness and Irishness in New Zealand since 1840* (Manchester, 2011), 153, 212.
4. John Stenhouse, 'Religion and Society—Church Adherence and Attendance, 1840–1920', in *Te Ara—the Encyclopedia of New Zealand* (2012), www.teara.govt.nz.
5. Martin Holland and Serena Kelly, 'Britain, Europe and New Zealand—Trade', *Te Ara—The Encyclopedia of New Zealand* (2012), www.teara.govt.nz.
6. Philippa Mein Smith, *A Concise History of New Zealand* (Cambridge, 2012), *passim*.
7. Mein Smith, *Concise History*, 129. Ian McGibbon, 'First World War', in McGibbon (ed.), OCNZMH, 174.
8. McGibbon, 'Gallipoli', in McGibbon (ed.), OCNZMH, 194.
9. Ashley Gould, 'Maori and the First World War', in McGibbon (ed.), OCNZMH, 297.
10. Christopher Pugsley, *The Anzac Experience: New Zealand, Australia and Empire in the First World War* (Auckland, 2004), 103–9.
11. Rhys Crawley, *Climax at Gallipoli: The Failure of the August Offensive* (Norman, OK, 2013).
12. Ron Palenski, 'Malcolm Ross: A New Zealand Failure in the Great War', *Australian Historical Studies* 39/1 (2008), 19–35; Ian McGibbon, '"Something of Them Is Here Recorded": Official History in New Zealand', in Jeffrey Grey (ed.), *The Last Word? Essays on Official History in the United States and British Commonwealth* (Westport, CT, 2003), http://nzetc.victoria.ac.nz/tm/schol arly/tei-McGSome.html.
13. The New Zealand official history of Gallipoli is Major Fred Waite, *The New Zealanders at Gallipoli* (Auckland, 1921).
14. 'Anzac Day', *Wanganui Chronicle*, 24 January 1916.
15. 'Untitled', *Otago Daily Times*, 15 March 1916.

16. 'Anzac Day', *Otago Daily Times*, 26 April 1916.
17. 'Soldiers at Dinner', *New Zealand Herald*, 26 April 1916.
18. 'Complimentary Concert', *New Zealand Herald*, 26 April 1916.
19. 'Returned Soldiers', *Auckland Star*, 26 April 1917; 'Soldiers Honoured', *New Zealand Herald*, 26 April 1918.
20. 'Patriotic Society', *Evening Post*, 14 April 1916.
21. 'Today's Celebrations', *Evening Post*, 25 April 1916.
22. 'Duty's Call', *Evening Post*, 26 April 1916.
23. Maureen Sharpe, 'Anzac Day in New Zealand: 1916 to 1939', *New Zealand Journal of History* 15/2 (1981), 97.
24. 'At the Synagogue', *Auckland Star*, 25 April 1916.
25. 'Prime Minister's Speech', *Evening Post*, 26 April 1916.
26. 'Anzac Day', *Otago Daily Times*, 25 April 1916.
27. 'Gallipoli Day', *Evening Post*, 29 April 1916.
28. Mercutio, 'Local Gossip', *New Zealand Herald*, 6 May 1916.
29. Scott Worthy, 'A Debt of Honour: New Zealanders' First Anzac Days', *New Zealand Journal of History* 36/2 (2002), 191–2.
30. 'Why Not? United Religious Service', *Auckland Star*, 25 March 1916.
31. 'Soldiers and Anzac Day', *Auckland Star*, 5 April 1916.
32. 'In the Dominion', *Hawera and Normanby Star*, 26 April 1916.
33. 'Local and General', *Evening Post*, 15 April 1916.
34. 'Honour the Heroes', *Evening Post*, 24 April 1916.
35. 'Local and General News', *Hawera and Normanby Star*, 17 April 1916.
36. 'Macmahon's Theatre', *Evening Post*, 25 April 1916.
37. 'Verdun', *Otago Daily Times*, 18 April 1916.
38. 'Letters to the Editor', *Otago Daily Times*, 19 April 1916.
39. 'Today's Celebrations', *Evening Post*, 25 April 1916.
40. 'Anzac Day', *Otago Daily Times*, 26 April 1916.
41. 'Anzac Day', *New Zealand Herald*, 26 April 1916.
42. 'Anzac Day', *Otago Daily Times*, 25 April 1916.
43. Tohunga, 'The Soldiers' Day', *New Zealand Herald*, 21 April 1917.
44. 'Anzac Day', *Hawera and Normanby Star*, 25 April 1918.
45. 'Anzac Day, 1919', *New Zealand Herald*, 25 April 1919.
46. Vivien, 'Topics of the Hour', *New Zealand Herald*, 22 April 1922.
47. 'Patriotism of the Profiteers', *NZ Truth*, 17 May 1919.
48. 'Anzac Day', *Otago Daily Times*, 13 November 1920.
49. '"As a Sunday"', *Evening Post*, 28 September 1921; 'Legislative Council: Anzac Day Bill', *Hawera and Normanby Star*, 27 January 1922.
50. Sapper, 'Anzac Day', *The Observer*, 24 January 1920.
51. 'Observance of Anzac Day', *New Zealand Herald*, 30 May 1918.
52. e.g. Okaiawa Sports Club, 'Okaiawa', *Hawera and Normanby Star*, 25 March 1919.
53. 'Anzac Day', *Ashburton Guardian*, 24 April 1919.

54. 'Anzac Day', *Auckland Star*, 25 April 1919.
55. Stephen Clarke, 'Return, Repatriation, Remembrance and the Returned Soldiers' Association 1916–1922', in John Crawford and Ian McGibbon (eds), *New Zealand's Great War: New Zealand, the Allies and the First World War* (Auckland, 2007), 167.
56. 'Anzac Day: Be Solemn!', *The Observer*, 26 April 1919.
57. Clarke, 'Return, Repatriation, Remembrance', 178.
58. Helen Robinson, 'Remembering the Past, Thinking of the Present: Historic Commemorations in New Zealand and Northern Ireland, 1940–1990' (PhD thesis, University of Auckland, 2009), 123.
59. Sharpe, 'Anzac Day in New Zealand', 99.
60. Mark David Sheftall, *Altered Memories of the Great War: Divergent Narratives of Britain, Australia, New Zealand and Canada* (London, 2009), 141.
61. Sharpe, 'Anzac Day in New Zealand', 106.
62. 'The Anzac Spirit', *Northern Advocate*, 24 April 1925.
63. 'Anzac Day', *Auckland Star*, 24 April 1925.
64. 'Impressive Memorial Service', *Evening Post*, 27 April 1925.
65. 'The Anzac Spirit', *Northern Advocate*, 24 April 1925.
66. Sharpe, 'Anzac Day in New Zealand', 104.
67. Worthy, 'A Debt of Honour', 195, 193.
68. McGibbon, 'Anzac Day', in McGibbon (ed.), *OCNZMH*, 28.
69. Jock Phillips, 'The Collinson and Cunningham Painting', in Fiona McKergow and Kerry Taylor (eds), *Te Hao Nui—The Great Catch* (Auckland, 2011), 128–33.
70. 'Too Mournful?', *Auckland Star*, 18 April 1929.
71. 'No Funeral March', *Auckland Star*, 29 March 1934.
72. 'Anzac Day Observance', *Auckland Star*, 5 February 1936.
73. 'Anzac Day Observance', *Evening Post*, 8 February 1936.
74. 'Returned Men Vote', *Evening Post*, 29 April 1936.
75. 'Too Gloomy', *Auckland Star*, 5 May 1938 1938.
76. 'Dawn Parade', *Auckland Star*, 27 April 1936.
77. 'Dawn Parades', *Evening Post*, 12 May 1938.
78. 'Dawn Parade', *Auckland Star*, 19 May 1938.
79. 'Anzac Day', *Auckland Star*, 24 March 1939.
80. Robinson, 'Remembering the Past, Thinking of the Present', 125.
81. 'Usual Service', *Auckland Star*, 12 March 1942.
82. 'Australia Drops Anzac Day', *Evening Post*, 11 March 1942.
83. Helen Robinson, 'Lest We Forget? The Fading of New Zealand War Commemorations, 1946–1966', *New Zealand Journal of History* 44/1 (2010), 76–7, 84.
84. Robinson, 'Remembering the Past, Thinking of the Present', 132.
85. 'Record Dawn Parade Marks 40th Anzac Anniversary', *Auckland Star*, 26 April 1955.
86. George Davis, 'Turkey's Engagement with Anzac Day, 1948–2000', *War & Society* 28/2 (2009), 133–61.

87. 'Valour and Bungling and Defeat', *New Zealand Herald*, 24 April 1965.
88. Robinson, 'Lest We Forget?', 78.
89. Roberto Rabel, 'Vietnam War', in McGibbon (ed.), *OCNZMH*, 563.
90. George Frederick Davis, 'Anzac Day Meanings and Memories: New Zealand, Australian and Turkish Perspectives on a Day of Commemoration in the Twentieth Century' (PhD, University of Otago, 2008), 234–44; Robinson, 'Remembering the Past, Thinking of the Present', 252–8.
91. Davis, 'Anzac Day Meanings and Memories', 240, 244.
92. Robinson, 'Remembering the Past, Thinking of the Present', 273, 74–83, 147.
93. Jock Phillips, 'Of Verandahs and Fish and Chips and Footie on Saturday Afternoon: Reflections on 100 Years of New Zealand Historiography', *New Zealand Journal of History* 24/2 (1990), 128–9.
94. Jock Phillips, 'The Quiet Western Front: The Great War and New Zealand Memory', in Santanu Das (ed.), *Race, Empire and First World War Writing* (Cambridge, 2011), 245.
95. Christopher Pugsley, *Gallipoli: The New Zealand Story* (Auckland, 1984).
96. Maurice Shadbolt, *Voices of Gallipoli* (Auckland, 1988), 114.
97. Robinson, 'Remembering the Past, Thinking of the Present', 275.
98. James Bennett, 'Man Alone and Men Together: Maurice Shadbolt, William Malone and Chunuk Bair', *Journal of New Zealand Studies*, NS 13 (2012), 52, 55–6. As Bennett notes, Pugsley suggests Malone was killed by New Zealand artillery.
99. W. David McIntyre, 'Anzus', in McGibbon (ed.), *OCNZMH*, 31–2.
100. Mein Smith, *Concise History*, 232–3.
101. McGibbon, 'Anzac Day' in McGibbon (ed.), *OCNZMH*, 29.
102. 'Relevance of Anzac Day', *Auckland Star*, 26 April 1985.
103. *Gallipoli Revisited: Anzac Cove 1990* (Canberra: National Film and Sound Archive, 1990), 54845.
104. Davis, 'Anzac Day Meanings and Memories', 261.
105. Graham Hucker, 'A Determination to Remember: Helen Clark and New Zealand's Military Heritage', *Journal of Arts Management, Law and Society* 40 (2010), 108, 113.
106. Helen Clark, 'Address at Dawn Service, Anzac Day' (2005), http://www.beehive.govt.nz/speech/address-dawn-service-anzac-day.
107. 'Chunuk Bair' (2000), http://www.beehive.govt.nz/speech/chunuk-bair.
108. 'Address at New Zealand 90th Anniversary Commemorations', at Chunuk Bair (2005), http://www.beehive.govt.nz/release/address-new-zealand-90th-anniversary-commemorations.
109. 'Address at Memorial Service for Unknown Warrior', http://www.beehive.govt.nz/node/21453/2004.
110. 'Mission and Objectives', http://ww100.govt.nz/about/about-WW100/mission-and-objectives#.U4c0UygkR2A.

111. Ian Allen, 'Anzac Bigger National Day, Says King', *Marlborough Express*, 7 February 2012.

112. Glyn Harper (ed.), *Letters from Gallipoli: New Zealand Soldiers Write Home* (Auckland, 2011), 245.

113. Jim McKay, '"We Didn't Want to Do a Dial-a-Haka": Performing New Zealand Nationhood in Turkey', *Journal of Sport & Tourism* 18/2 (2013), 117–35.

Chapter 7

1. 'Up the Republic', *Connacht Tribune*, 23 September 1933.

2. 'Attempt to Ban Picture', *Connacht Tribune*, 9 September 1933.

3. 'IRA Resolution', *Connacht Tribune*, 16 December 1933.

4. 'A War Film', *Irish Press*, 10 November 1931.

5. 'Motion Picture', *Sunday Independent*, 4 October 1931.

6. 'Mr Anthony Asquith', *The Times*, 22 February 1968.

7. Jane Leonard, 'The Twinge of Memory: Armistice Day and Remembrance Sunday in Dublin since 1919', in Richard English and Graham Walker (eds), *Unionism in Modern Ireland: New Perspectives on Politics and Culture* (Basingstoke, 1996), 99–114.

8. 'The Message of Anzac Day', *The Times*, 26 April 1916; 'In Honour of Anzac', *The Times*, 26 April 1916. E. M. Andrews, *The Anzac Illusion: Anglo-Australian Relations During World War One* (Cambridge, 1993), 88; Eric Andrews, '25 April 1916: First Anzac Day in Australia and Britain', *Journal of the Australian War Memorial* 23 (October 1993), 16.

9. 'Heroes from Gallipoli', *Manchester Guardian*, 15 March 1916.

10. 'The Links of Empire', *Manchester Guardian*, 27 May 1916.

11. 'The March of the Anzacs', *The Times*, 26 April 1919.

12. Geoffrey Moorhouse, *Hell's Foundations: A Town, Its Myths and Gallipoli* (London, 1992), 147–8. For an evocative description of the service and parade held each year, see ibid. 155–9.

13. 'Lancashire Heroes', *Manchester Guardian*, 8 May 1916.

14. 'Gallipoli Day', *Manchester Guardian*, 3 June 1916; 'Gallipoli Day', *Manchester Guardian*, 6 June 1916.

15. 'Gallipoli—July 12th, 1915', *Southern Reporter*, 13 July 1916.

16. From an Anzac Correspondent, 'Gallipoli Day', *The Times*, 25 April 1917.

17. 'The Glory of Gallipoli', *The Times*, 26 April 1917.

18. 'Anzac Day', *The Times*, 26 April 1918.

19. Bruce Scates et al., '"Such a Great Space of Water between Us": Anzac Day in Britain, 1916–1939', *Australian Historical Studies* 45 (2014), 228.

20. 'Anzac Day', *Manchester Guardian*, 26 April 1919. See also e.g. '"Anzac Day" in Manchester', *Manchester Guardian*, 26 April 1923; 'Anzac Day', *Manchester Guardian*, 26 April 1924.

21. 'Untitled', *Irish Times*, 22 August 1918; 'The Gallipoli Decoration', *Manchester Guardian*, 5 September 1918; 'The Gallipoli Medal', *Manchester Guardian*, 5 September 1918; 'The Gallipoli Decoration', *Manchester Guardian*, 12 September 1918.

22. e.g. 'Gallipoli Landing of Lancashire Fusiliers', *Manchester Guardian*, 25 April 1927; 'Lancashire Landing at Gallipoli', *Manchester Guardian*, 23 April 1928.

23. 'Bristol and the War', *Western Daily Press*, 7 April 1917; 'Gallipoli Day', *Western Daily Press*, 17 April 1919.

24. 'Gallipoli Day', *Western Daily Press*, 24 April 1924.

25. 'Gallipoli Landing', *Western Morning News and Mercury*, 26 April 1921; 'Anzac Day', *The Times*, 26 April 1922; 'The Landing at Gallipoli', *Western Morning News and Mercury*, 26 April 1928.

26. Robin Prior, *Gallipoli: The End of the Myth* (New Haven, 2009), 242. This figure was reached by Prior by including non-battle casualties, extrapolated from the British non-battle casualties which were double their battle casualties.

27. 'The Withdrawal from Anzac', *The Times*, 22 December 1915; 'German View of the British Withdrawal', *Newcastle Daily Journal*, 22 December 1916; 'Announcement in the Reichstag', *Coventry Evening Telegraph*, 22 December 1915.

28. 'Book of the Week: Gallipoli from the Other Side', *Aberdeen Press and Journal*, 8 October 1928; 'Lord Allenby's "Last Crusade": Lloyd George on a Great Romance', *Courier and Advertiser*, 4 December 1928.

29. J. M. Winter, *Sites of Memory, Sites of Mourning: The Great War in European Cultural History* (Cambridge, 1995).

30. Jenny Macleod, *Reconsidering Gallipoli* (Manchester, 2004).

31. John Masefield, *Gallipoli* (London, 1916).

32. Patrick Shaw-Stewart, 'I saw a man this morning', in David Childs and Vivien Whelpton (eds), *British and Irish Poets of the Gallipoli Campaign, 1915: Heirs to Achilles* (London, 2011), 30.

33. Childs and Whelpton (eds), *British and Irish Poets of the Gallipoli Campaign, 1915*, 53.

34. A. P. Herbert, *The Secret Battle* (London, 1919).

35. 'Gallipoli: Heroes' Imperishable Fame', *Devon and Exeter Gazette*, 11 January 1916.

36. 'Victory Slips from Britain's Hands', *Evening Telegraph and Post*, 10 January 1916.

37. 'The Turks Again Outwitted', *Manchester Guardian*, 10 January 1916.

38. 'The Message of Anzac Day', *The Times*, 26 April 1916.

39. 'In Honour of Anzac', *The Times*, 26 April 1916.

40. 'June the Fourth', *Manchester Guardian*, 5 June 1916.

41. 'Gallipoli Day', *The Times*, 25 April 1917.

42. 'The Dardanelles Report', *The Times*, 9 March 1917.

43. 'Dardanelles Report', *The Times*, 19 November 1919.

44. 'A Contrast', *Western Times*, 9 March 1917.

45. 'Our Daily Survey: The Greatest Failure', *Evening Express*, 9 March 1917.
46. 'Responsibility for the Dardanelles Muddle', *Evening Telegraph*, 8 March 1917.
47. 'The Dardanelles Report', *The Times*, 9 March 1917.
48. 'The Dardanelles Debate', *Manchester Guardian*, 21 March 1917.
49. 'Mr. Churchill', *Manchester Guardian*, 20 July 1917.
50. 'The War against Turkey', *Manchester Guardian*, 11 April 1916.
51. 'The Crisis of the War and of the Ministry', *Manchester Guardian*, 4 December 1916.
52. 'Constantinople and the Straits', *Manchester Guardian*, 4 August 1921.
53. 'Gallipoli', *The Times*, 17 May 1920; 'Gallipoli Diary', *Exeter and Plymouth Gazette*, 17 May 1920.
54. General Sir Ian Hamilton, *Ian Hamilton's Despatches from the Dardanelles* (London, 1917).
55. Ernest Raymond, *Please You, Draw Near: Autobiography 1922–1968* (London, 1969), 69.
56. Ernest Raymond, *Tell England: A Study in a Generation* (London, 1922), 273.
57. 'Mr Churchill's Book—I', *The Times*, 8 October 1923, the serialization concluded on 27 October 1923; 'Agonising Story of the Dardanelles from Within', *Hull Daily Mail*, 30 October 1923.
58. Sir Winston Churchill, *The World Crisis 1915* (London, 1923); Robin Prior, *Churchill's 'World Crisis' as History* (London, 1983).
59. 'Mr. Churchill and "The World Crisis"', *The Observer*, 4 November 1923.
60. C. F. Aspinall-Oglander, *Military Operations, Gallipoli*, 2 vols (London, 1929 and 1932). For an extended discussion see, Macleod, *Reconsidering Gallipoli*, 81–93.
61. 'Gallipoli Landing', *Manchester Guardian*, 27 April 1925.
62. Scates et al., '"Such a Great Space of Water between Us"', 235.
63. Moorhouse, *Hell's Foundations*, 149.
64. 'Gallipoli', *Southern Reporter*, 9 July 1925.
65. Keith Grieves, 'Remembering an Ill-Fated Venture: The Fourth Battalion, Royal Sussex Regiment at Suvla Bay and Its Legacy, 1915–1939', in Jenny Macleod (ed.), *Gallipoli: Making History* (London, 2004), 119.
66. 'The Lesson of Gallipoli', *Colonist*, 24 April 1917; 'The Anzacs', *The Argus*, 25 April 1917.
67. Keith Jeffery, 'Gallipoli and Ireland', in Macleod (ed.), *Gallipoli: Making History*, 105; Stuart Ward, 'Parallel Lives, Poles Apart: Commemorating Gallipoli in Ireland and Australia', in John Horne and Edward Madigan (eds), *Towards Commemoration: Ireland in War and Revolution, 1912–1923* (Dublin, 2013), 32.
68. 'The Irish in Gallipoli', *Irish Times*, 10 November 1917.
69. 'Tribute to the Irish in Gallipoli', *Freeman's Journal*, 27 March 1916.
70. 'Gallipoli Hero', *Freeman's Journal*, 8 May 1918.
71. Childs and Whelpton (eds), *British and Irish Poets of the Gallipoli Campaign*, 1915, 42, 48, 52.
72. 'Tragic Chapter's End', *Irish Independent*, 10 January 1916.

73. 'Gallipoli', *Irish Times*, 15 April 1916.
74. 'Ireland and the War', *Irish Times*, 19 August 1916.
75. 'Gallipoli', *Irish Times*, 28 October 1916.
76. 'Gallipoli of Ireland', *Irish Independent*, 18 July 1917.
77. 'The Irish at the Front', *Irish Times*, 25 March 1916. A number of other accounts of the Irish at Gallipoli were published during the war, including Michael MacDonagh, *The Irish at the Front* (London, 1916); S. Parnell Kerr, *What the Irish Regiments Have Done* (London, 1916); and Henry Hanna, *The Pals at Suvla Bay* (Dublin, 1917).
78. 'Fair Play', 'The Forgotten 10th (Irish) Division' *Irish Times*, 10 January 1917.
79. E.B., 'The Tenth (Irish) Division' *Irish Times*, 16 April 1919.
80. 'The Irish in Gallipoli', *Irish Times*, 10 November 1917.
81. Ward, 'Parallel Lives, Poles Apart', 34.
82. 'Matters of Moment: Suppressed Despatches', *Irish Independent*, 5 May 1919.
83. Philip Orr, *Field of Bones: An Irish Division at Gallipoli* (Dublin, 2006), 188.
84. 'Anzac Day', *Irish Times*, 25 April 1916.
85. 'Anzac Day', *Irish Times*, 25 April 1919.
86. 'Anzac Day', *Irish Times*, 26 April 1922.
87. 'The Royal Munster Fusiliers', *Irish Times*, 25 April 1927; 'Gallipoli Heroes', *Irish Times*, 30 April 1928; 'The Munster Fusiliers', *Irish Times*, 14 April 1932.
88. 'The Landing at Gallipoli', *Irish Times*, 22 April 1929.
89. 'Pro Patria', *Irish Times*, 26 April 1927.
90. 'Twenty Years Ago', *Irish Times*, 26 April 1935.
91. Orr, *Field of Bones*, 219; Fergus A. D'Arcy, *Remembering the War Dead: British Commonwealth and International War Graves in Ireland since 1914* (Dublin, 2007), 181.
92. 'The National War Memorial', *Irish Times*, 15 December 1931.
93. Keith Jeffery, *Ireland and the Great War* (Cambridge, 2000), 131–2.
94. Jeffery, *Ireland and the Great War*, 134.
95. D'Arcy, *Remembering the War Dead*, 347, 352, 354.
96. 'Anzac Day Mass in Dublin', *Irish Times*, 26 April 1947. He also conducted the service in 1948, 1953, and 1954 at least.
97. 'Anzac Day', *Irish Times*, 21 April 1955.
98. 'Anzac Day Mass in Dublin Church', *Irish Times*, 26 April 1956.
99. 'What's on This Weekend', *Irish Times*, 25 April 1987.
100. 'April 25 Is Just Another Day', *News*, 8 April 1965. Press Cuttings: Anzac Day 1965–6 (ACT and Overseas), Australian War Memorial, Canberra.
101. Jay Winter and Antoine Prost, *The Great War in History: Debates and Controversies, 1914 to the Present* (Cambridge, 2005), 16–17.
102. For further details, see Macleod, *Reconsidering Gallipoli*, 213–16.
103. 'Battles Long Ago', *The Times*, 22 April 1965.
104. A. J. P. Taylor, 'Last Word on Gallipoli', *The Observer*, 25 April 1965.
105. Francis McManus, 'Death at the Dardanelles', *Irish Press*, 24 April 1965.

106. Dan Todman, *The Great War: Myth and Memory* (London, 2005), 214, 208–19; Todman, 'The Internet and the Remembrance of the Two World Wars', *RUSI Journal* 155/5 (2010), 76–81; Helen B. McCartney, 'The First World War Soldier and His Contemporary Image in Britain', *International Affairs* 90/2 (2014), 304–5.

107. *The Gallipolian* 66 (1991), personal correspondence with Keith Edmonds, membership secretary.

108. Keith Jeffery, 'Irish Varieties of Great War Commemoration', in John Horne and Edward Madigan (eds), *Towards Commemoration: Ireland in War and Revolution, 1912–1923* (Dublin, 2013), 121; Tom Burke, 'Rediscovery and Reconciliation: The Royal Dublin Fusiliers', in Horne and Madigan (eds), *Towards Commemoration*, 98–104.

109. Dominic Bryan and Stuart Ward, 'The "Deficit of Remembrance": The Great War Revival in Australia and Ireland', in Katie Holmes and Stuart Ward (eds), *Exhuming Passions: The Pressure of the Past in Ireland and Australia* (Dublin, 2011), 182.

110. 'Speech by An Taoiseach, Mr Bertie Ahern T.D.', *Blue Cap: Journal of the Royal Dublin Fusiliers* 8 (June 2001), 28.

111. 'Remember Them', *The Times*, 26 April 2000. New Zealand's last veteran, Doug Dibley, died in 1997. Australia gave state funerals to Ted Matthews in 1997 and to Alec Campbell in 2002. Matthews was the last surviving Anzac to have landed on the peninsula on 25 April 1915; Campbell was the last surviving veteran of the campaign.

112. *The Gallipolian* 93 (2000), 19.

113. *The Gallipolian* 96 (2001), 2.

114. Alrewas, Bury, Chepstow, Dersingham, Edinburgh, Eltham, Gourock, Hawick, Manchester, Isle of Wight, Nottingham, Pimperne, Stretton-on-Dunsmore, Cork, and Limerick. This list is based on information distributed each year by the Gallipoli Association through its publication *The Gallipolian* and its website, www.gallipoli-assocation.org.

115. *The Gallipolian* 109 (2005), 41.

116. *The Gallipolian* 101 (2003), 26.

117. Aberdeen, Arbroath, Baverstock, Brockenhurst, Cannock Chase, Cambridge, Cardigan, Chilworth, Codford St Mary, Dallachy Strike Wing, Glenavy, Harefield, Leighterton, Manchester, Morecambe, Norfolk/Suffolk (alternating years), Northampton, Oxford, Peterborough, Portsmouth, Salisbury, St Albans, Sutton Veny, Weymouth, Warrington, Walton-on-Thames, Dublin.

Chapter 8

1. 'Davutoğlu: 12 yıl Sonra Cihan Devletiyiz', *Takvim*, 25 April 2011, http://www.takvim.com.tr/Siyaset/2011/04/25/davutoglu-12-yil-sonra-cihan-devletiyiz.

2. Erik J. Zürcher, *The Young Turk Legacy and Nation Building: From the Ottoman Empire to Ataturk's Turkey* (London, 2010), 108.

3. Mehmet Besikçi, *The Ottoman Mobilization of Manpower in the First World War: Between Voluntarism and Resistance* (Leiden, 2012), 113.

4. Zürcher, *The Young Turk Legacy*, 139.

5. Haluk Oral, *Gallipoli 1915: Through Turkish Eyes*, trans. Amy Spangler (Istanbul, 2007), 351.

6. Ayhan Aktar, 'The Making of Mustafa Kemal's Saga at Gallipoli', unpublished conference paper, February 2014.

7. The poem was published for instance in the *Harb Mecmuası* (War Magazine), 11 (August 1916); *Tasvir-i Efkar* (Representation of Opinions), 1857 (8 September 1916); *Servet-i Fünun* (Wealth of Knowledge), 1317 (14 September 1916). The poem was also published in the special issue of the *Yeni Mecmua* (New Magazine), 'Çanakkale Nüsha-yı Fevkalâdesi' (Gallipoli extraordinary issue), in 1918.

8. For the names of the attendees, see İbrahim Alaettin [Gövsa]'s introduction to the 2nd edition of the *Çanakkale İzleri* (1932 [1st pub. 1922]). See also Erol Köroğlu, *Ottoman Propaganda and Turkish Identity* (London, 2007), 82–3. After the 1934 surname law, some of them changed their names as follows: Ali Canip [Yöntem], Celal Sahir [Erozan], Enis Behiç [Koryürek], Hakkı Süha [Gezgin], Hamdullah Suphi [Tanrıöver], Hıfzı Tevfik [Gönensay], Orhan Seyfi [Orhon], Selahattin, Mehmed Emin [Yurdakul], İbrahim Alaeddin [Gövsa], Ahmed Yekta [Madran], Çallı İbrahim and Nazmi Ziya [Güran].

9. BOA, MF.MKT, 1210/28, 22 Şaban 1333 [5 July 1915], cited in Mustafa Selçuk, 'Birinci Dünya Savaşı Sürecinde Harbiye Nezareti'nin "Çanakkale Kahramanlığını Yaşatma" Amaçlı Faaliyetleri', *Avrasya İncelemeleri Dergisi*, I/2 (2012), 200.

10. 'Çanakkale: 5-18 Mart 1331–1915, Yeni Mecmua'nın Fevkalade Nüshası', *Yeni Mecmua*, 1918. 1331 is the year according to the *Rumi* calendar, equivalent to 1915 in the Gregorian calendar.

11. 'Çanakkale 5-18 Mart Zaferi' (1918)', *Donanma Mecmuası*, 1918.

12. Esra Özyürek, 'Introduction', in Özyürek (ed.), *The Politics of Public Memory in Turkey* (New York, 2007), 4–6.

13. Uğur Ümit Üngör, *The Making of Modern Turkey: Nation and State in Eastern Anatolia, 1913–1950* (Oxford, 2011), 241–2, 219, 231.

14. Zürcher, *The Young Turk Legacy*, 49–50.

15. Ayhan Aktar, '"Turkification" Policies in the Early Republican Era', in Catharina Dufft (ed.), *Turkish Literature and Cultural Memory: 'Multiculturalism' as a Literary Theme after 1980* (Wiesbaden, 2009), 31.

16. Üngör, *The Making of Modern Turkey*, 228–9.

17. Fatma Müge Göçek, 'Reading Genocide: Turkish Historiography on 1915,' in Ronald Grigor Suny, Fatma Müge Göçek, and Norman M. Naimark (eds),

A *Question of Genocide: Armenians and Turks at the End of the Ottoman Empire* (Oxford, 2011), 43.

18. Aktar, 'The Making of Mustafa Kemal's Saga at Gallipoli'.

19. Gavin D. Brockett, *How Happy to Call Oneself a Turk: Provincial Newspapers and the Negotiation of a Muslim National Identity* (Austin, 2011), 76.

20. Ayhan Aktar, 'Debating the Armenian Massacres in the Last Ottoman Parliament, November–December 1918', *History Workshop Journal*, 64 (2007), 248, 257.

21. Zürcher, *The Young Turk Legacy*, 202.

22. Erik Jan Zürcher, 'Renewal and Silence: Postwar Unionist and Kemalist Rhetoric on the Armenian Genocide,' in Suny et al. (eds), *A Question of Genocide*, 312.

23. Zürcher, *The Young Turk Legacy*, 168.

24. Winston Churchill, *The World Crisis 1915* (London, 1923), 322, 444–51.

25. Ayhan Aktar, 'The Making of Mustafa Kemal's Saga at Gallipoli', 14–20; Jenny Macleod, *Reconsidering Gallipoli* (Manchester, 2004), 85.

26. 'Çanakkale Atatürk Demektir', *Cumhuriyet*, 1 August 1936.

27. Quoted in Edward J. Erickson, *Ordered to Die: A History of the Ottoman Army in the First World War* (Westport, CT, 2001), 83.

28. Telegrams were also sometimes sent to other leaders during the ceremonies. İsmet İnönü, Kazım Karabekir, and Fevzi Pasha also received telegrams in which they were praised, but Atatürk's telegrams were always the longest and the most flattering.

29. Şehitlikleri İmar Cemiyeti was founded in 1926, but the online archives for *Cumhuriyet* only begin in 1930, so it has been impossible to examine their activities before this date.

30. 'Çanakkale Şehitlerini Ziyaret', *Cumhuriyet*, 4 July 1933.

31. Janda Gooding, *Gallipoli Revisited: In the Footsteps of Charles Bean and the Australian Historical Mission* (Melbourne, Victoria and London, 2009), 65–6.

32. Bart Ziino, *A Distant Grief: Australians, War Graves and the Great War* (Crawley, WA, 2007), 66.

33. John McQuilton, 'Gallipoli as Contested Commemorative Space', in Macleod (ed.), *Gallipoli: Making History*, 152.

34. Ziino, *A Distant Grief*, 77.

35. Anatol Shmelev, 'Gallipoli to Golgotha: Remembering the Internment of the Russian White Army at Gallipoli, 1920–1923', in Jenny Macleod (ed.), *Defeat and Memory: Cultural Histories of Military Defeat in the Modern Era* (Basingstoke, 2008), 195–6.

36. 'Anzac Day', *The Argus*, 13 August 1920.

37. Ziino, *A Distant Grief*, 70–4.

38. David W. Lloyd, *Battlefield Tourism: Pilgrimage and the Commemoration of the Great War in Britain, Australia and Canada* (Oxford, 1998), 97.

39. 'Bir "Anzak"ın Gazi Hz. Yolladığı Mektup', *Cumhuriyet*, 25 April 1934.

40. 'Anzac Day in Australia', *The Times*, 26 April 1934; 'Former Enemy', *Sydney Morning Herald*, 26 April 1934. See also George Davis, 'Turkey's Engagement with Anzac Day, 1948–2000', *War & Society* 28/2 (2009), 136–7 which notes a similar message to a Brisbane newspaper in 1931, and the obscurity of the 1934 speech.

41. Uluğ İğdemir, *Atatürk Ve Anzaklar = Atatürk and the Anzacs* (Ankara, 1978).

42. 'Gallipoli Pilgrims', *Sydney Morning Herald*, 5 May 1934; Our Own Correspondent, 'British Veterans Visit Gallipoli', *The Times*, 3 May 1934; Stanton Hope, *Gallipoli Revisited: An Account of the Duchess of Richmond Pilgrimage-Cruise* (London, 1934). Hope's very full account of the pilgrimage refers to an exchange of radio messages with 'the Gazi', but not to any further speeches on the peninsula.

43. Adrian Jones, 'A Note on Ataturk's Words About Gallipoli', *History Australia* 2/1 (2004), 10–17,8.

44. Abidin Daver, 'Anzaklar', *Cumhuriyet*, 14 February 1940.

45. 'Çanakkale Harbinin Yıldönümü Bugün Mısır'da Kutlanacak', *Cumhuriyet*, 25 April 1940.

46. 'Çanakkale Zaferi', *Cumhuriyet*, 19 March 1946.

47. Erik Zürcher, *Turkey: A Modern History* (London, 1993; repr. 2004), 234.

48. '18 Mart Çanakkale Şehitleri İhtifali', *Milliyet*, 10 March 1952.

49. 'Çanakkale Zaferinin 37nci Yıldönümü Dün Muhteşem Bir Törenle Kutlandı', *Milliyet*, 19 March 1952.

50. 'Çanakkale Zaferinin Yıldönümü', *Milliyet*, 18 March 1953.

51. 'Çanakkale Gezisi Geniş Bir İlgiyle Karşılandı', *Milliyet*, 09 July 1952.

52. 'Çanakkale'de Yapılacak Gezi Yarın Başlıyor', *Milliyet*, 9 August 1952.

53. 'Çanakkale'ye Yapılacak Büyük Gezi', *Milliyet*, 10 August 1952.

54. Davis, 'Turkey's Engagement with Anzac Day, 1948–2000', 139.

55. 'Anafartalar Muharebesinin Yıldönümü', *Milliyet*, 10 August 1953.

56. 'Kısa Haberler', *Milliyet*, 16 March 1952.

57. '18 Mart Çanakkale Zaferinin Yıldönümü', *Milliyet*, 18 March 1952.

58. 'Çanakkale'de Bir Mehmetçik Abidesi Yaptırılıyor', *Milliyet*, 5 May 1952.

59. 'Çanakkale Zaferimizi Bugün Kutluyoruz', *Milliyet*, 18 March 1958.

60. 'Çanakkale Zaferimizi Bugün Kutluyoruz', *Milliyet*, 18 March 1958; 'Çanakkale Zaferinin 43: Yıldönümü', *Milliyet*, 17 March 1958.

61. 'Çanakkale Zaferinin 44: Yıldönümü', *Milliyet*, 18 March 1959.

62. 'Çanakkale Zaferinin 45: Yılı Dün Kutlandı', *Milliyet*, 19 March 1960.

63. The memorial was officially opened on 21 August 1960.

64. 'Gallipoli to-Day', *West Australian*, 20 December 1920.

65. 'Çanakkale Abidesi', *Milliyet*, 17 April 1954.

66. 'Çanakkale Anıtı İçin 1 Milyona Daha İhtiyaç Var', *Milliyet*, 28 July 1957; 'Çanakkale Abidesi', *Milliyet*, 30 July 1957.

67. 'Yeni Zelandalı Eski Muhariplerin Çanakkale Abidesi İçin Teberuu', *Milliyet*, 20 September 1958.

68. 'Çanakkale'den Yeni Zelanda'ya Taş Götürülecek', *Milliyet*, 9 January 1959.
69. 'Çanakkalede Türk Şehitliği: Yok', *Milliyet*, 9 February 1958.
70. Refi' Cevad Ulunay, 'Çanakkale Abidesi Hakkında', *Milliyet*, 10 April 1961.
71. 'Çanakkale Abidesi İçin 500 Bin Fidan Ayrıldı', *Milliyet*, 18 April 1963.
72. 'Çanakkale Şehitler Abidesinin Çatı İnşaatı Tamamlandı', *Milliyet*, 19 February 1966.
73. John McQuilton, 'Gallipoli as Contested Commemorative Space', in Macleod (ed.), *Gallipoli: Making History*, 150–4.
74. 'Çanakkale Zaferinin Yıldönümü Kutlandı', *Milliyet*, 19 March 1964.
75. 'Çanakkale Zaferi Kutlandı', *Milliyet*, 19 March 1967.
76. 'Çanakkale Zaferi Kutlandı', *Milliyet*, 19 March 1968.
77. Sabri Sayarı, 'Political Violence and Terrorism in Turkey: 1976–1980: A Retrospective Analysis', *Terrorism and Political Violence*, 22 (2010), 198–215.
78. 'Çanakkale Zaferi Törenlerine Katılan Ülkü Ocakları Birliği Üyeleri Hükümet Aleyhinde Gösteri Yaptı', *Cumhuriyet*, 19 March 1974.
79. According to the news published in the Milliyet for the 62nd anniversary, neither this year, nor last year's ceremonies allowed students to participate. *Milliyet*, 19 March 1977.
80. 'Çanakkale Zaferi'nin 64. Yıldönümü Kutlandı', *Milliyet*, 19 March 1979.
81. For further details, see Sinan Ciddi, *Kemalism in Turkish Politics: The Republican People's Party, Secularism and Nationalism* (London, 2009).
82. For further information, see Esra Özyürek, 'Miniaturizing Atatürk: Privatisation of State Imagery and Ideology of the State in Turkey', *American Ethnologist* 3/3 (2004), 374–91.
83. 'Çanakkale Özel', *Milliyet*, 11 August 1981.
84. 'Çanakkale Zaferi'nin 67: Yıldönümü Kutlandı', *Milliyet*, 19 March 1982.
85. 'Çanakkale Kahramanları Anıldı', *Milliyet*, 19 March 1987.
86. 'Atatürk'süz Çanakkale Programına Tepki Yağıyor', *Milliyet*, 20 March 1988.
87. 'Unutulmayan Destan', *Milliyet*, 19 March 1986.
88. 'Çanakkale Zaferi'nin 73: Yıldönümü Kutlandı', *Milliyet*, 19 March 1988.
89. Zürcher, *Turkey: A Modern History*, 288.
90. Jenny White, *Muslim Nationalism and the New Turks* (Princeton, 2013), 160.
91. 'Kraliçe Çanakkale'de Şehitliği Ziyaret Etti', *Milliyet*, 23 October 1971.
92. Davis, 'Turkey's Engagement with Anzac Day, 1948–2000', pp. 151 n. 47, 143.
93. 'Son 'Gazi'lerin Sefaleti', *Milliyet*, 18 March 1990; 'Sahipsiz Gaziler', *Milliyet*, 29 April 1990.
94. 'Çanakkale Gazilerine Madalya', *Milliyet*, 15 November 1990, 15.
95. Özyürek, 'Introduction', 2.
96. 'Çanakkale'ye Altın Madalya', *Milliyet*, 18 March 1994.
97. Zürcher, *Turkey: A Modern History*, 264.
98. 'Ülkenin Hiçbir Taşı Geçilmez', *Cumhuriyet*, 19 March 1993.
99. 'Çiller: Tanzimat Ruhuyla İleri', *Milliyet*, 19 March 1997.
100. 'Haber Merkezi', *Milliyet*, 19 March 1997.

101. Önder Yılmaz, 'Çanakkale', *Milliyet*, 19 March 1998.
102. I am grateful to Professor Gencer Özcan for his guidance on this point.
103. Meliha Benli Altunışık and Özlem Tür Kavli, *Turkey: Challenges of Continuity and Change* (New York, 2004), 88.
104. Hakan Yılmaz, *Secularism and Muslim Democracy in Turkey* (Cambridge, 2009), 95.
105. 'Baba: Enver Paşa Kahramandır', *Milliyet*, 30 July 1996.
106. Hew Strachan, *The First World War* (Oxford, 2001), 728.
107. Norman Stone, *Turkey: A Short History* (London, 2010), 145.
108. Enis Oznuk and Yasin Canakci, 'Sarikamiş Martyrs Honored After 89 Years', *Today's Zaman*, 6 July 2003, www.todayszaman.com/news-2322-sarikamis-martyrs-honored-after-89-years.html.
109. 'Erdoğan'ın Çanakkale Zaferi'nin Yıldönümü Konuşması', *Cumhuriyet*, 18 March 2010, www.cumhuriyet.com.tr/haber/diger/128186/Erdogan_in_Canakkale_Zaferi_nin_yildonumu_konusmasi.html.
110. Alphan Eseli, *The Long Way Home* (2013),www.thelongwayhome-themovie.com/en/. The film won two awards at the Montreal World Film Festival 2013: 'The Long Way Home', International Movie Database, www.imdb.com/title/tt2608766/awards?ref_=tt_ql_4 (accessed 24 February 2014).
111. Ali Aslan Kılıç, '100th anniversaries become a source of inspiration from economy to culture', *Today's Zaman*, 13 January 2013, www.todayszaman.com/news-303929-100th-anniversaries-become-a-source-of-inspiration-from-economy-to-culture.html.
112. Başak İnce, *Citizenship and Identity in Turkey: From Atatürk's Republic to the Present Day* (London and New York, 2012), 173–4.
113. Cengiz Güneş, 'Political Reconciliation in Turkey: Challenges and Prospects', in Cengiz Güneş and Welat Zeydanlıoğlu (eds), *The Kurdish Question in Turkey: New Perspectives on Violence, Representation, and Reconciliation* (London and New York, 2014), 270.
114. 'Başbakan Erdoğan: Çanakkale'yi Anlamayan Bizi Anlayamaz', 18 March 2013, http://haber.stargazete.com/politika/basbakan-erdogan-can akkaleyi-anlamayan-turkiyeyi-anlayamaz/haber-736998.
115. These billboards were installed at 10 different locations in the city centre of Diyarbakır, in 2013 for the 98th anniversary of 18 March, http://www.cnnturk.com/2013/turkiye/03/16/diyarbakira.kurtce.canakkale.afisi/700487.0/index.html.
116. Üngör, *The Making of Modern Turkey*, 110–22.
117. Peter Balakian, *The Burning Tigris: The Armenian Genocide and America's Response* (New York, 2003), loc 3163.
118. Kramer, *Dynamic of Destruction*, 146–50, quote at 148.
119. Gülay Türkmen-Dervişoğlu, 'Coming to Terms with a Difficult Past: The Trauma of the Assassination of Hrant Dink and Its Repercussions on Turkish National Identity', *Nations and Nationalism* 19/4 (2013), 679.

120. Taner Akçam, *A Shameful Act: The Armenian Genocide and the Question of Turkish Responsibility*, trans. Paul Bessemer (New York, 2006), 126.

121. 'Ottoman Armenians during the Decline of the Ottoman Empire: Issues of Scientific Responsibility and Democracy', Istanbul, September 2005.

122. Yavuz Baydar, 'Turkey Should Acknowledge the Armenian Genocide', *Al-Monitor*, 25 April 2013.

123. Aktar, 'Debating the Armenian Massacres', 245–6.

124. Erdem Suna, 'Turk Who Defied Official History Wins Nobel Prize', *The Times*, 13 October 2006.

125. Erdem Suna, 'Three Held after "Anti-Turkish" Editor Killed', *The Times*, 20 January 2007.

126. Giles Whittell, 'History That Dares to Speak Its Name', *The Times*, 18 August 2007; 'Human Rights Court Rules Turkey Cannot Criminalize Genocide Recognition', www.asbarez.com, 25 October 2011; Case 27520/07 *Case of Altuğ Taner Akçam v Turkey*, [2011] ECHR.

127. Helen Davidson, 'Gallipoli Service: O'Farrell Attacks Turkish Threat to Bar NSW MPs', *The Guardian*, 22 August 2013; Ben MacIntyre, 'What's the Turkish for Genocide?', *The Times*, 18 June 2005.

128. Anthony Browne, 'Turkey Will Not Apologise for Armenian Genocide', *The Times*, 15 December 2004; Ben MacIntyre and Chris Harris, 'Brutality, Poverty and Religion Stand between Turkey', *The Times*, 30 September 2005.

129. John D. McKinnon and Marc Champion, 'White House Puts Brakes on Armenia Vote', *Wall Street Journal (Online)*, 6 March 2010; Bronwen Maddox, 'Stirring up the Past, Jeopardising the Future', *The Times*, 17 October 2007.

130. 'Erdogan'in Canakkale Zaferi'nin Yildonumu Konusmasi', *Cumhuriyet*, 18 March 2010.

131. Lauren McMah, 'NSW Parliament Formally Recognises Assyrian Genocide', *Daily Telegraph* (accessed 12 March 2014). South Australia has also recognized the genocide. See also Colin Tatz, 'Turkey, the Armenian Genocide and the Politics of Memory', *The Conversation*, 19 December 2013, www.theconversation.com.

132. Michael Brissenden, 'Turkey Threatens to Ban MPs from Gallipoli Centenary over Genocide Vote', *ABC News*, 21 August 2013.

133. 'Armenian Genocide Denier Justin McCarthy to Speak at Parliament House', *ABC News*, 16 November 2013.

134. Anonymous, 'Davutoglu Says Turkey Sees Australia as "Strategic Partner"', *BBC Monitoring European*, 26 April 2011.

135. Robert Fisk, 'The Armenian Hero Turkey Would Prefer to Forget', *The Independent*, 12 May 2013.

136. Turhan Gürkan, 'Çanakkale Zaferinin Tek Filmi', *Cumhuriyet*, 15 March 1995.

137. Catherine Simpson, 'From Ruthless Foe to National Friend: Turkey, Gallipoli and Australian Nationalism', *Media International Australia*, 137 (2010), 63. The

film was criticized by the Turkish ambassador to Australia for omitting Atatürk's 1934 address.

138. Sinan Çetin, *Çanakkale Çocukları* (Canakkale Children) (2012); Yeşim Sezgin, *Çanakkale 1915* (2012); Kemal Uzun and Serdar Akar, *Çanakkale Yolun Sonu* (Canakkale End of the Line) (2013). Plus the aforementioned *The Long Way Home* about Sarikamiş, and a movie about the Turkish War of Independence: Altan Dönmez, *Taş Mektep* (Stone School) (2013). Phil Hoad, 'Will Platoon of Gallipoli Films Give Turkish Audiences Battle Fatigue?', *The Guardian*, 9 May 2013.

139. Musaffer Altunay and Önder Deligöz, 'Visitors Gather for Dawn Service While Turks Re-Enact 57th Regiment March', *Today's Zaman*, 26 April 2007.

140. 'Başbakan Erdoğan: Çanakkale'yi Anlamayan Bizi Anlayamaz', 18 March 2013.

141. Brad West, 'Dialogical Memorialization, International Travel and the Public Sphere: A Cultural Sociology of Commemoration and Tourism at the First World War Gallipoli Battlefields', *Tourist Studies* 10/3 (2010), 209–25.

Chapter 9

1. Untitled note, *Le Figaro*, 25 April 1920; 'Le Pèlerinage des poilus d'Orient aux Dardanelles', *Le Matin*, 13 April 1930; untitled note, *Le Matin*, 25 April 1935. See also David Dutton, '"Docile Supernumerary": A French Perspective on Gallipoli', in Jenny Macleod (ed.), *Gallipoli: Making History* (London, 2004).

2. Philip Orr, *Field of Bones: An Irish Division at Gallipoli* (Dublin, 2006), 230.

3. 'Erdoğan: Bizleri Savaşta Karşı Karşıya Getirdiler', 10 January 2013, http://www.haber7.com/dis-politika/haber/975579-erdogan-bizleri-savasta-karsi-karsiya-getirdiler.

4. I look forward to Peter Stanley's *Die in Battle; Do not Despair: Indians on Gallipoli, 1915* (Solihull, forthcoming).

5. John Henry Patterson, *With the Zionists in Gallipoli* (London, 1916); Martin Sugarman, 'The Zion Muleteers of Gallipoli, March 1915–May 1916', *Jewish Historical Studies* 36 (2001), 113–39.

6. John Gallishaw, *Trenching at Gallipoli: A Personal Narrative of a Newfoundlander with the Ill-Fated Dardanelles Expedition* (New York, 1916); P. Whitney Lackenbauer, 'War, Memory, and the Newfoundland Regiment at Gallipoli', *Newfoundland Studies* 14/2 (1999), 176–214.

SELECT BIBLIOGRAPHY

Archival Sources

Anzac Day Press cuttings files for 1965 and 1990, Australian War Memorial, Canberra.

Ashmead-Bartlett Papers, A1583, Mitchell Library, Sydney.

Bean Papers, AWM 38 3 DRL 6673, Australian War Memorial, Canberra.

Dardanelles Commission proceedings, CAB 19/33, The National Archives, London.

Gallipoli Revisited: Anzac Cove 1990, TV Documentary, 54845, National Film and Sound Archive, Canberra.

Newspapers

1. *Australia*

Newspapers marked with an asterisk were found in trove.nla.gov.au.

*The Argus (Melbourne, Victoria)
The Australian (national)
Australian Times (www.australiantimes.co.uk)
*Brisbane Courier (Queensland)
*Camperdown Chronicle (Victoria)
*Canberra Times (Australian Capital Territory)
Daily Advertiser (Canberra, Australian Capital Territory)
*Examiner (Launceston, Tasmania)
*Kadina and Wallaroo Times (South Australia)
*The Mercury (Hobart, Tasmania)
*Morning Bulletin (Rockhampton, Queensland)
News (Adelaide, South Australia)
*Referee (Sydney, New South Wales)
*The Register (Adelaide, South Australia)
*Riverine Herald (Echuca, Victoria; Moama, New South Wales)
The Sun-Herald (New South Wales)
Sunday Canberra Times (Australian Capital Territory)
*Sunday Times (Perth, Western Australia)
*Sydney Morning Herald (New South Wales)

Warwick Examiner and Times (Queensland)
West Australian (Perth, Western Australia)

2. Britain
Newspapers marked with an asterisk were found in www.britishnewspaperarchive.
co.uk

Aberdeen Press and Journal
Courier and Advertiser (Dundee)
Coventry Evening Telegraph
Devon and Exeter Gazette
Evening Express (Aberdeen)
Exeter and Plymouth Gazette
The *Guardian* (www.theguardian.com)
Hull Daily Mail
Impartial Reporter (Enniskillen)
Manchester Guardian (Guardian and Observer Archive 1791–2003, www.proquest.
com)
Newcastle Daily Journal
The *Observer* (Guardian and Observer Archive 1791–2003, www.proquest.com)
Southern Reporter (Selkirk)
The *Times* (The Times Digital Archive, www.galegroup.com)
Western Daily Press (Bristol)
Western Morning News and Mercury (Plymouth, Devon)
Western Times (Exeter, Devon)

3. Ireland
Newspapers marked with an asterisk were found in www.irishnewsarchive.com

Connacht Tribune (Galway)
Freeman's Journal (Dublin)
Impartial Reporter (Enniskillen)
Irish Independent (Dublin)
Irish Press (Dublin)
Irish Times (Dublin) (www.irishtimes.com/archive)

4. France
Le Figaro (gallica.bnf.fr)
Le Matin (gallica.bnf.fr)

5. New Zealand
All newspapers were found in www.paperspast.natlib.govt.nz. Post-Second World
War editions of the *Auckland Star* and the *New Zealand Herald* were consulted in
the Alexander Turnbull Library, National Library of New Zealand, Wellington.

Ashburton Guardian (Canterbury)
Auckland Star (Auckland)
Evening Post (Wellington)
Hawera and Normanby Star (Taranaki)
Marlborough Express (Marlborough)
New Zealand Herald (Auckland)
Northern Advocate (Northland)
NZ Truth (National)
Observer (Auckland)
Otago Daily Times (Otago)
Wanganui Chronicle (Manawatu-Wanganui)

6. Turkey
Al-Monitor (www.al-monitor.com)
Cumhuriyet (www.cumhuriyetarsivi.com)
Donanma Mecmuası
Harb Mecmuasi
Milliyet (http://gazetearsivi.milliyet.com.tr)
Servet-i Fünun
Takvim (www.takvim.com.tr)
Taraf
Tasvir-i Efkar
Today's Zaman (www.todayszaman.com)
Yeni Mecmua

Other Online Sources

ABC News, ABC Local Radio, Australia, www.abc.net.au
Asbarez newspaper (Fresno, California), www.Asbarez.com
BBC Monitoring European, www.Proquest.com
Case of Altuğa Taner Akçam V. Turkey (2011), European Court of Human Rights, http://hudoc.echr.coe.int
'Decade of Centenaries', Ireland, www.decadeofcentenaries.com
Hansard, Parliament of Australia, www.aph.gov.au
'historypunk' blog by Jo Hawkins, www.historypunk.com
'PM Transcripts', speeches by Australian prime ministers, pmtranscripts.dpmc.gov.au
The official website of the New Zealand Government, www.beehive.govt.nz
Wall Street Journal (Online), www.Proquest.com
WW100 New Zealand, New Zealand government centenary website, www.ww100.govt.nz

Other Primary Sources

Australian War Memorial Annual Report 1991–1992 (1992)
Australian War Memorial Annual Report 2005–2006 (2006)
Australian War Memorial Annual Report 2007–2008 (2008)
The Gallipolian: The Journal of the Gallipoli Association

Secondary Sources

Adam-Smith, Patsy, *The Anzacs* (London: Hamish Hamilton, 1978).

Akçam, Taner, *A Shameful Act: The Armenian Genocide and the Question of Turkish Responsibility*, trans. Paul Bessemer (New York: Metropolitan Books/Henry Holt and Company, 2006).

Aksakal, Mustafa, *The Ottoman Road to War in 1914: The Ottoman Empire and the First World War* (Cambridge: Cambridge University Press, 2008).

Aktar, Ayhan, 'Debating the Armenian Massacres in the Last Ottoman Parliament, November–December 1918', *History Workshop Journal* 64 (2007), 240–70.

Aktar, Ayan, '"Turkification" Policies in the Early Republican Era', in Catharina Dufft (ed.), *Turkish Literature and Cultural Memory: 'Multiculturalism' as a Literary Theme after 1980* (Wiesbaden: Harrassowitz Verlag, 2009), 29–62.

Aktar, Ayhan, 'The Making of Mustafa Kemal's Saga at Gallipoli: A British Imperial and Turkish Joint Operation, 1921–1932', paper presented at 'The British Empire and the Great War—Colonial Societies/ Cultural Responses', Nanyang Technological University, Singapore, 19–22 February 2014.

Alomes, Stephen, *A Nation at Last? The Changing Character of Australian Nationalism 1880–1988* (North Ryde, NSW: Angus & Robertson, 1988).

Altunışık, Meliha Benli, and Özlem Tür Kavli, *Turkey: Challenges of Continuity and Change* (New York: Routledge, 2004),

Andrews, E. M., 'Bean and Bullecourt: Weaknesses and Strengths of the Official History of Australia in the First World War', *Revue internationale d'histoire militaire* 72 (August 1990), 25–47.

Andrews, E. M., *The Anzac Illusion: Anglo-Australian Relations During World War One* (Cambridge: Cambridge University Press, 1993).

Andrews, Eric, '25 April 1916: First Anzac Day in Australia and Britain', *Journal of the Australian War Memorial* 23 (October 1993), 13–20.

'Australian Commonwealth Horse', in Peter Dennis, Jeffrey Grey, Ewan Morris, Robin Prior, and with John Connor (eds), *The Oxford Companion to Australian Military History* (Melbourne: Oxford University Press, 1995), 65.

Ashmead-Bartlett, Ellis, *Ashmead-Bartlett's Despatches from the Dardanelles: An Epic of Heroism* (London: George Newnes, 1916).

Ashmead-Bartlett, Ellis, *The Uncensored Dardanelles* (London: Hutchinson & Co., 1928).

Aspinall-Oglander, C. F., *Military Operations, Gallipoli*, History of the Great War Based on Official Documents, ed. Sir James Edmonds, 2 vols (London: William Heinemann Ltd, 1929–32).

Balakian, Peter, *The Burning Tigris: The Armenian Genocide and America's Response* (New York: HarperCollins e-books, 2003).

Bean, C. E. W., *The Story of Anzac from the Outbreak of War to the End of the First Phase of the Gallipoli Campaign, May 4, 1915*, The Official History of Australia in the War of 1914–1918, i, ed. Robert O'Neill (12 vols, St Lucia, Queensland: University of Queensland Press, 1921; repr. 1981).

Bean, C. E. W., *The Story of Anzac: From 4 May, 1915 to the Evacuation of the Gallipoli Peninsula*, The Official History of Australia in the War of 1914–1918, ii, ed. Robert O'Neill (12 vols, St Lucia, Queensland: University of Queensland Press, 1924; repr. 1981).

Bean, C. E. W., 'The Writing of the Australian Official History of the Great War: Sources, Methods and Some Conclusions', *Royal Australian Historical Society, Journal and Proceedings* 24/2 (1938), 85–112.

Bean, C. E. W., 'The Technique of a Contemporary War Historian', *Historical Studies, Australia and New Zealand* 2, 6 (1942), 65–79.

Beaumont, Joan, 'The Anzac Legend', in Beaumont (ed.), *Australia's War, 1914–1918* (St Leonards, NSW: Allen & Unwin, 1995).

Bennett, James, 'Man Alone and Men Together: Maurice Shadbolt, William Malone and Chunuk Bair', *Journal of New Zealand Studies*, NS 13 (2012), 46–61.

Besikçi, Mehmet, *The Ottoman Mobilization of Manpower in the First World War: Between Voluntarism and Resistance* (Leiden: Brill, 2012).

Blair, Dale, *Dinkum Diggers: An Australian Battalion at War* (Melbourne: Melbourne University Press, 1997).

Broadbent, Harvey, *Gallipoli: The Fatal Shore* (Camberwell, Victoria: Penguin, 2005).

Brockett, Gavin D., *How Happy to Call Oneself a Turk: Provincial Newspapers and the Negotiation of a Muslim National Identity* (Austin: University of Texas Press, 2011).

Bromfield, Nicholas, 'Welcome Home: Reconciliation, Vietnam Veterans, and the Reconstruction of Anzac under the Hawke Government', http://www.auspsa.org.au/sites/default/files/welcome_home_nicholas_bromfield.pdf.

Brown, James, *Anzac's Long Shadow: The Cost of Our National Obsession* (Collingwood, Victoria: Redback, 2014).

Bryan, Dominic, and Ward, Stuart, 'The "Deficit of Remembrance": The Great War Revival in Australia and Ireland', in Katie Holmes and Stuart Ward (eds), *Exhuming Passions: The Pressure of the Past in Ireland and Australia* (Dublin: Irish Academic Press, 2011), 163–86.

Burke, Tom, 'Rediscovery and Reconcilisation: The Royal Dublin Fusiliers', in John Horne and Edward Madigan (eds), *Towards Commemoration: Ireland in War and Revolution, 1912–1923* (Dublin: Royal Irish Academy, 2013), 98–104.

Butler, Janet, 'Nursing Gallipoli: Identity and the Challenge of Experience', *Journal of Australian Studies* 27/78 (2003), 47–57.

Carlyon, Les, *Gallipoli* (Sydney: Pan Macmillan Australia, 2001).

Cash, Damien, 'Seymour, Alan', in William H. Wilde, Joy Hooton, and Barry Andrews (eds), *The Oxford Companion to Australian Literature* (Oxford: Oxford University Press, 1994).

Cassar, George, 'Kitchener at the War Office', in Hugh Cecil and P. H. Liddle (eds), *Facing Armageddon: The First World War Experienced* (London: Cooper, 1996), 37–50.

Cassar, George H., *The French and the Dardanelles: A Study of Failure in the Conduct of War* (London: George Allen & Unwin, 1971).

Chasseaud, P., and Doyle, P., *Grasping Gallipoli: Terrain, Maps and Failure at the Dardanelles, 1915* (Staplehurst: Spellmount, 2005).

Childs, David, and Whelpton, Vivien (eds), *British and Irish Poets of the Gallipoli Campaign, 1915: Heirs to Achilles* (London: Cecil Woolf Publishers, 2011).

Churchill, Sir Winston, *The World Crisis 1915* (London: Thornton Butterworth, 1923).

Ciddi, Sinan, *Kemalism in Turkish Politics: The Republican People's Party, Secularism and Nationalism* (London: Routledge, 2009).

Clarke, Stephen, 'Return, Repatriation, Remembrance and the Returned Soldiers' Association 1916–1922', in John Crawford and Ian McGibbon (eds), *New Zealand's Great War: New Zealand, the Allies and the First World War* (Auckland: Exisle, 2007), 157–81.

Cochrane, Peter, *Simpson and the Donkey: The Making of a Legend* (Melbourne: Melbourne University Press, 1992).

Commonwealth Bureau of Census and Statistics, *Year Book of the Commonwealth of Australia*, 6 (Melbourne: McCarron, Bird & Co, 1913).

Connor, John, *Anzac and Empire: George Foster Pearce and the Foundations of Australian Defence* (Cambridge: Cambridge University Press, 2011).

Crawley, Rhys, *Climax at Gallipoli: The Failure of the August Offensive* (Norman, OK: University of Oklahoma Press, 2013).

Crotty, Martin, 'The Anzac Citizen: Towards a History of the RSL', *Australian Journal of Politics and History* 53/2 (2007), 183–93.

Crotty, Martin, and Melrose, Craig, 'Anzac Day, Brisbane, Australia: Triumphalism, Mourning and Politics in Interwar Commemoration', *Round Table* 96/393 (2007), 679–92.

Curran, James, *The Power of Speech: Australian Prime Ministers Defining the National Image* (Carlton, Victoria: Melbourne University Press, 2004).

Curran, James, and Ward, Stuart, *The Unknown Nation: Australia after Empire* (Carlton, Victoria: Melbourne University Press, 2010).

D'Arcy, Fergus A., *Remembering the War Dead: British Commonwealth and International War Graves in Ireland since 1914* (Dublin: The Stationery Office, 2007).

Damousi, Joy, 'Private Loss, Public Mourning: Motherhood, Memory and Grief in Australia During the Inter-War Years', *Women's History Review* 8/2 (1999), 365–78.

Davis, George, 'Turkey's Engagement with Anzac Day, 1948–2000', *War & Society* 28/2 (2009), 133–61.

Davis, George Frederick, 'Anzac Day Meanings and Memories: New Zealand, Australian and Turkish Perspectives on a Day of Commemoration in the Twentieth Century' (PhD, University of Otago, 2008).

Dawson, Danny, 'The "Harefield/Anzac Bond": A Study of Thanksgiving and Remembrance in an English Village', *Family and Community History* 9/1 (2006), 27–42.

Donaldson, Carina, and Lake, Marilyn, 'Whatever Happened to the Anti-War Movement?', in Lake, Henry Reynolds, Mark McKenna, and Joy Damousi (eds), *What's Wrong with Anzac?* (Sydney: University of New South Wales Press, 2010), 71–93.

Dutton, David, '"Docile Supernumerary": A French Perspective on Gallipoli', in Jenny Macleod (ed.), *Gallipoli: Making History* (London: Frank Cass, 2004), 86–97.

Dutton, Philip, '"More Vivid Than the Written Word": Ellis Ashmead-Bartlett's Film, *With the Dardanelles Expedition* (1915)', *Historical Journal of Film, Radio and Television* 24/2 (2004), 205–22.

Erickson, Edward J., *Ordered to Die: A History of the Ottoman Army in the First World War* (Westport, CT: Praeger Publishers, 2000).

Erickson, Edward J., 'Strength Against Weakness: Ottoman Military Effectiveness at Gallipoli, 1915', *Journal of Military History* 65 (2001), 981–1102.

Erickson, Edward J., *Gallipoli: The Ottoman Campaign* (Barnsley: Pen & Sword Military, 2010).

Fewster, Kevin, 'Ellis Ashmead-Bartlett and the Making of the Anzac Legend', *Journal of Australian Studies* 10 (June 1982), 17–30.

Foster, Roy, 'Remembering 1798', in Ian McBride (ed.), *History and Memory in Modern Ireland* (Cambridge: Cambridge University Press, 2001), 67–94.

Gallishaw, John, *Trenching at Gallipoli: A Personal Narrative of a Newfoundlander with the Ill-Fated Dardanelles Expedition* (New York: Century Co., 1916).

Gammage, Bill, *The Broken Years: Australian Soldiers in the Great War* (Ringwood, Victoria: Penguin Books Australia, 1974; repr. 1975).

Gare, Deborah, 'Britishness in Recent Australian Historiography', *Historical Journal* 43/4 (2000), 1145–55.

Gilbert, Martin, *Winston S. Churchill*, iii. *Companion Part I, Documents July 1914–April 1915* (London: Heinemann, 1972).

Gilbert, Martin, *The Straits of War: Gallipoli Remembered* (Stroud: Sutton, 2000).

Göncü, Gürsel, and Aldoğan, Şahin, *Siperin Ardı Vatan* (Istanbul: MB Yayinlari, 2006).

Gooding, Janda, *Gallipoli Revisited: In the Footsteps of Charles Bean and the Australian Historical Mission* (Melbourne, Victoria and London: Hardie Grant, 2009).

Gould, Ashley, 'Maori and the First World War', in Ian McGibbon (ed.), *The Oxford Companion to New Zealand Military History* (Auckland: Oxford University Press, 2000), 296–9.

[Gövsa] İbrahim Alaettin. *Çanakkale İzleri: Anafartalar'ın Müebbet Kahramanına* (Istanbul: Semih Lütfi-Sühulet Kütüphanesi, 1922; repr. 1932).

Gregory, Adrian, *The Silence of Memory: Armistice Day 1919–1946* (Oxford: Berg Publishers, 1994).

Grieves, Keith, 'Remembering an Ill-Fated Venture: The Fourth Battalion, Royal Sussex Regiment at Suvla Bay and Its Legacy, 1915–1939', in Jenny Macleod (ed.), *Gallipoli: Making History* (London: Cass, 2004), 110–24.

Güneş, Cengiz, 'Political Reconciliation in Turkey: Challenges and Prospects', in Cengiz Güneş and Welat Zeydanlıoğlu (eds), *The Kurdish Question in Turkey: New Perspectives on Violence, Representation, and Reconciliation* (London and New York: Routledge, Taylor & Francis Group, 2014),

Hamilton, Sir I. S. M., *Gallipoli Diary*, 2 vols (London: Edward Arnold, 1920).

Harper, Glyn (ed.), *Letters from Gallipoli: New Zealand Soldiers Write Home* (Auckland: Auckland University Press, 2011).

Harrison, Mark, *The Medical War: British Military Medicine in the First World War* (Oxford: Oxford University Press, 2010).

Hart, Peter, *Gallipoli* (London: Profile Books, 2011).

Herbert, A. P., *The Secret Battle* (London: Methuen & Co., 1919).

Herbert, Aubrey, *Mons, Anzac and Kut* (London: Edward Arnold, 1919).

Hickey, Michael, *Gallipoli* (London: John Murray, 1995).

Holland, Martin, and Kelly, Serena, 'Britain, Europe and New Zealand—Trade', in *Te Ara—The Encyclopedia of New Zealand* (2012), www.teara.govt.nz/en.

Hope, Stanton, *Gallipoli Revisited: An Account of the Duchess of Richmond Pilgrimage-Cruise* (London: the author, 1934).

Horne, John, 'Our War, Our History', in Horne (ed.), *Our War: Ireland and the Great War. The 2008 Thomas Davis Lecture Series* (Dublin: Royal Irish Academy, 2008), 1–34.

Hucker, Graham, 'A Determination to Remember: Helen Clark and New Zealand's Military Heritage', *Journal of Arts Management, Law and Society* 40 (2010), 105–18.

İğdemir, Uluğ, *Atatürk Ve Anzaklar = Atatürk and the Anzacs* (Ankara: Türk Tarih Kurumu Basımevi, 1978).

İnce, Başak, *Citizenship and Identity in Turkey: From Atatürk's Republic to the Present Day* (London and New York: I. B. Tauris, 2012).

Inglis, Kenneth Stanley, 'The Australians at Gallipoli, 2 Pts'. *Historical Studies* 14 (1970), 219–30 and 361–75.

Inglis, K. S., *Sacred Places: War Memorials in the Australian Landscape* (Carlton South, Victoria: Melbourne University Press, 1998; repr. 2001).

Inglis, K. S. 'Anzac, the Substitute Religion', in Craig Wilcox (ed.), *Observing Australia 1959 to 1999* (Carlton South, Victoria: Melbourne University Press, 1999), 61–70.

Inglis, Ken, 'Men, Women, and War Memorials: Anzac Australia', in *Anzac Remembered: Selected Writings of K. S. Inglis*, ed. John Lack (Melbourne: RMIT Publishing, 2001).

Inglis, K. S. 'Return to Gallipoli', in *Anzac Remembered: Selected Writings of K. S. Inglis*, ed. John Lack (Melbourne: RMIT Publishing, 2001).

Jeffery, Keith, *Ireland and the Great War* (Cambridge: Cambridge University Press, 2000).

Jeffery, Keith, 'Gallipoli and Ireland', in Jenny Macleod (ed.), *Gallipoli: Making History* (London: Frank Cass, 2004), 98–109.

Jeffery, Keith, 'Irish Varieties of Great War Commemoration', in John Horne and Edward Madigan (eds), *Towards Commemoration: Ireland in War and Revolution, 1912–1923* (Dublin: Royal Irish Academy, 2013), 117–25.

Jones, Adrian, 'A Note on Ataturk's Words About Gallipoli', *History Australia* 2/1 (2004), 10-11–10-19.

Kannengiesser Pasha, Hans, *The Campaign in Gallipoli*, trans. Major C. J. P. Ball DSO MC (Late RA) (London: Hutchinson & Co. Ltd, 1928).

Kent, D. A., 'The *Anzac Book* and the Anzac Legend: C. E. W. Bean as Editor and Image-Maker', *Historical Studies* 21/84 (April 1985), 376–90.

Köroğlu, Erol, *Ottoman Propaganda and Turkish Identity* (London: I. B. Tauris, 2007).

Kramer, Alan, *Dynamic of Destruction: Culture and Mass Killing in the First World War* (Oxford: Oxford University Press, 2007).

Lackenbauer, P. Whitney, 'War, Memory, and the Newfoundland Regiment at Gallipoli'. *Newfoundland Studies* 14/2 (Fall 1999), 176–214.

Lake, Marilyn, 'Introduction: What Have You Done for Your Country?', in Lake, Henry Reynolds, Mark McKenna, and Joy Damousi (eds), *What's Wrong with Anzac? The Militarisation of Australian History* (Sydney: University of New South Wales Press, 2010), 1–23.

Lee, John, *A Soldier's Life: General Sir Ian Hamilton 1853–1947* (London: Macmillan, 2000).

Leonard, Jane, 'The Twinge of Memory: Armistice Day and Remembrance Sunday in Dublin since 1919', in Richard English and Graham Walker (eds), *Unionism in Modern Ireland: New Perspectives on Politics and Culture* (Basingstoke: Palgrave Macmillan, 1996), 99–114.

Liman von Sanders, Otto, *Five Years in Turkey*, trans. US Army (Retired) Colonel Carl Reichmann (Annapolis, MD: The United States Naval Institute, 1920; repr. 1927).

Liman von Sanders, Marshal Otto, 'The Campaign of Gallipoli, 1915–1916', *Royal United Services Institution Journal* 67/465 (1922), 147–8.

Lloyd, David W., *Battlefield Tourism: Pilgrimage and the Commemoration of the Great War in Britain, Australia and Canada* (Oxford: Berg, 1998).

McCarthy, Angela, *Scottishness and Irishness in New Zealand since 1840* (Manchester: Manchester University Press, 2011).

McGibbon, Ian, 'Anzac', in McGibbon (ed.), *The Oxford Companion to New Zealand Military History* (Auckland: Oxford University Press, 2000), 27.

McGibbon, Ian, 'Anzac Day', in McGibbon (ed.), *The Oxford Companion to New Zealand Military History* (Auckland: Oxford University Press, 2000), 27–30.

McGibbon, Ian, 'First World War', in McGibbon(ed.), *The Oxford Companion to New Zealand Military History* (Auckland: Oxford University Press, 2000), 172–5.

McGibbon, Ian, 'Gallipoli', in McGibbon (ed.), *The Oxford Companion to New Zealand Military History* (Auckland: Oxford University Press, 2000), 190–8.

McGibbon, Ian, '"Something of Them Is Here Recorded": Official History in New Zealand', in Jeffrey Grey (ed.), *The Last Word? Essays on Official History in the United States and British Commonwealth* (Westport, CT: Praeger, 2003), http://nzetc.victoria.ac.nz/tm/scholarly/tei-McGSome.html.

Macintyre, Stuart, *The Oxford History of Australia*, iv. *The Succeeding Age 1901–1942*, The Oxford History of Australia, ed. Geoffrey Bolton (Melbourne: Oxford University Press, 1986; repr. 2001).

McIntyre, W. David, 'Anzus', in Ian McGibbon (ed.), *The Oxford Companion to New Zealand Military History* (Auckland: Oxford University Press, 2000), 31–2.

McKay, Jim, '"We Didn't Want to Do a Dial-a-Haka": Performing New Zealand Nationhood in Turkey', *Journal of Sport & Tourism* 18/2 (2013), 117–35.

McKenna, Mark, 'Anzac Day: How Did It Become Australia's National Day?', in Marilyn Lake, Henry Reynolds, Mark McKenna, and Joy Damousi (eds), *What's Wrong with Anzac?* (Sydney: University of New South Wales Press, 2010), 110–34.

McKenzie, Andrea, '"Our Common Colonial Voices": Canadian Nurses, Patient Relations, and Nation on Lemnos', in Joachim Bürgschwentner, Matthias Egger, and Gunda Barth-Scalmani (eds), *Other Fronts, Other Wars? First World War Studies on the Eve of the Centennial* (Leiden: Brill, 2014).

Mackenzie, Compton, *Gallipoli Memories* (London: Cassell & Co., 1929).

Macleod, Jenny, 'General Sir Ian Hamilton and the Dardanelles Commission', *War in History* 8/4 (2001), 418–41.

Macleod, Jenny, 'The Fall and Rise of Anzac Day: 1965 and 1990 Compared', *War & Society* 20/1 (May 2002), 149–68.

Macleod, Jenny, 'The British Heroic-Romantic Myth of Gallipoli', in Macleod (ed.), *Gallipoli: Making History* (London: Frank Cass, 2004), 73–85.

Macleod, Jenny, *Reconsidering Gallipoli* (Manchester: Manchester University Press, 2004).

Macleod, Jenny, 'Ellis Ashmead-Bartlett, War Correspondence and the First World War', in Y. T. McEwen and F. A. Fisken (eds), *War, Journalism and History: War Correspondents in the Two World Wars* (Bern: Peter Lang, 2012), 31–48.

Macleod, Jenny, 'Britishness and Commemoration: National Memorials to the First World War in Britain and Ireland', *Journal of Contemporary History* 48/4 (2013), 647–65.

McMeekin, Sean, *The Berlin–Baghdad Express: The Ottoman Empire and Germany's Bid for World Power, 1898–1918* (London: Allen Lane, 2010).

McQuilton, John, 'Gallipoli as Contested Commemorative Space', in Jenny Macleod (ed.), *Gallipoli: Making History* (London: Cass, 2004), 150–8.

Masefield, John, *Gallipoli* (London: W. Heinemann, 1916).

Mein Smith, Philippa, *A Concise History of New Zealand* (Cambridge: Cambridge University Press, 2012).

Moorehead, Alan, *Gallipoli* (London: Hamish Hamilton, 1956).

Moorhouse, Geoffrey, *Hell's Foundations: A Town, Its Myths and Gallipoli* (London: Hodder & Stoughton, 1992).

Morgenthau, Henry, *Ambassador Morgenthau's Story: A Personal Account of the Armenian Genocide* (New York: Cosimo Books, 1918; repr. 2010).

Moseley, S. A., *The Truth About a Journalist* (London: Pitman, 1935).

Moses, John A., 'The Struggle for Anzac Day 1916–1930 and the Role of the Brisbane Anzac Day Commemoration Committee', *Journal of the Royal Australian Historical Society* 88/1 (2002), 54–74.

Moses, John A., 'Gallipoli or Other Peoples' Wars Revisited: Sundry Reflections on Anzac: A Review Article', *Australian Journal of Politics and History* 57/3 (2011), 434–42.

Neilson, Keith, 'Kitchener, Horatio Herbert, Earl Kitchener of Khartoum (1850–1916)', in *Oxford Dictionary of National Biography* (Oxford: Oxford University Press, 2004).

Nevinson, Henry W., *Last Changes, Last Chances* (London: Nisbet and Co. Ltd, 1928).

Nora, Pierre, 'General Introduction: Between Memory and History', in Pierre Nora (ed.), *Realms of Memory: Rethinking the French Past*, i. *Conflicts and Divisions* (New York: Columbia University Press, 1996), 1–20.

Oral, Haluk, *Gallipoli 1915: Through Turkish Eyes*, trans. Amy Spangler (Istanbul: Türkiye İş Bankası Kültür Yayınları, 2007).

Orr, Philip, *Field of Bones: An Irish Division at Gallipoli* (Dublin: The Lilliput Press, 2006).

Özdemir, Hikmet, *The Ottoman Army 1914–1918: Disease and Death on the Battlefield* (Salt Lake City: University of Utah Press, 2008).

Özyürek, Esra, 'Miniaturizing Atatürk: Privatisation of State Imagery and Ideology of the State in Turkey', *American Ethnologist* 3/3 (2004), 374–91.

Özyürek, Esra, 'Introduction', in Özyürek (ed.), *The Politics of Public Memory in Turkey* (New York: Syracuse University Press, 2007), 1–15.

Palenski, Ron, 'Malcolm Ross: A New Zealand Failure in the Great War', *Australian Historical Studies* 39/1 (2008), 19–35.

Patterson, John Henry, *With the Zionists in Gallipoli* (London: Hutchinson & Co., 1916).

Pavils, J. G., *Anzac Day: The Undying Debt* (Adelaide: Lythrum Press, 2007).

Phillips, Jock, 'Of Verandahs and Fish and Chips and Footie on Saturday Afternoon: Reflections on 100 Years of New Zealand Historiography', *New Zealand Journal of History* 24/2 (1990), 118–34.

Phillips, Jock, 'The Collinson and Cunningham Painting', in Fiona McKergow and Kerry Taylor (eds), *Te Hao Nui—The Great Catch* (Auckland: Godwit Books, 2011), 128–33.

Phillips, Jock, 'The Quiet Western Front: The Great War and New Zealand Memory', in Santanu Das (ed.), *Race, Empire and First World War Writing* (Cambridge: Cambridge University Press, 2011), 231–48.

Prior, Robin, *Gallipoli: The End of the Myth* (New Haven: Yale University Press, 2009).

Pugsley, Christopher, *Gallipoli: The New Zealand Story* (Auckland: Hodder & Stoughton, 1984).

Pugsley, Christopher, *The Anzac Experience: New Zealand, Australia and Empire in the First World War* (Auckland: Reed Publishing, 2004).

Rabel, Roberto, 'Vietnam War', in Ian McGibbon (ed.), *The Oxford Companion to New Zealand Military History* (Auckland: Oxford University Press, 2000), 561–6.

Raymond, Ernest, *Tell England: A Study in a Generation* (London: Cassell & Co., 1922).

Rhodes James, Robert, *Gallipoli* (London: Pimlico, 1965; repr. 1999).

Robinson, Helen, 'Remembering the Past, Thinking of the Present: Historic Commemorations in New Zealand and Northern Ireland, 1940–1990' (PhD thesis, University of Auckland, 2009).

Robinson, Helen, 'Lest We Forget? The Fading of New Zealand War Commemorations, 1946–1966', *New Zealand Journal of History* 44/1 (2010), 76–91.

Rooney, Chris B., 'The International Significance of British Naval Missions to the Ottoman Empire, 1908–1914', *Middle Eastern Studies* 34/1 (1998), 1–29.

Ross, Jane, *The Myth of the Digger: The Australian Soldier in Two World Wars* (Sydney: Hale & Iremonger, 1985).

Rudenno, Victor, *Gallipoli: Attack from the Sea* (New Haven: Yale University Press, 2008).

Sayarı, Sabri, 'Political Violence and Terrorism in Turkey: 1976–1980: A Retrospective Analysis', *Terrorism and Political Violence* 22 (2010), 198–215.

Scates, Bruce C., *Return to Gallipoli: Walking the Battlefields of the Great War* (Cambridge: Cambridge University Press, 2006).

Scates, Bruce, Bongiorno, Frank, Wheatley, Rebecca, and James, Laura, '"Such a Great Space of Water between Us": Anzac Day in Britain, 1916–1939', *Australian Historical Studies* 45 (2014), 220–41.

Seal, Graham, '"...And in the Morning...": Adapting and Adopting the Dawn Service', *Journal of Australian Studies* 35/1 (2011), 49–63.

Shadbolt, Maurice, *Voices of Gallipoli* (Auckland: Hodder & Stoughton, 1988).

Sharpe, Maureen, 'Anzac Day in New Zealand: 1916 to 1939', *New Zealand Journal of History* 15/2 (1981), 97–114.

Sheftall, Mark David, *Altered Memories of the Great War: Divergent Narratives of Britain, Australia, New Zealand and Canada* (London: I. B. Tauris, 2009).

Shmelev, Anatol, 'Gallipoli to Golgotha: Remembering the Internment of the Russian White Army at Gallipoli, 1920–1923', in Jenny Macleod (ed.), *Defeat and Memory: Cultural Histories of Military Defeat in the Modern Era* (Basingstoke: Palgrave Macmillan, 2008), 195–213.

Simpson, Catherine, 'From Ruthless Foe to National Friend: Turkey, Gallipoli and Australian Nationalism', *Media International Australia*, 137 (2010), 58–66.

Stanley, Peter, 'Australia, India and Gallipoli: A Study in Contrasting National Consequences', (unpublished lecture), Australian War Memorial, 2004.

Stanley, Peter, *Quinn's Post: Anzac, Gallipoli* (Sydney: Allen & Unwin Australia, 2005).

Steel, Nigel, '"Heroic Sacrifice": The Sikh Regiment at Gallipoli, June 1915', Portraits of Courage lecture, Imperial War Museum North, 2004.

Steel, Nigel, and Hart, Peter, *Defeat at Gallipoli* (London: Pan, 2002).

Stenhouse, John, 'Religion and Society—Church Adherence and Attendance, 1840–1920', in *Te Ara—the Encyclopedia of New Zealand* (2012), www.teara.govt.nz/en.

Stone, Norman, *Turkey: A Short History* (London: Thames and Hudson, 2010).

Strachan, Hew, *The First World War* (Oxford: Oxford University Press, 2001).

Strachan, Hew, 'The First World War as a Global War', *First World War Studies* 1/1 (2010), 3–14.

Sugarman, Martin, 'The Zion Muleteers of Gallipoli, March 1915–May 1916', *Jewish Historical Studies* 36 (2001), 113–39.

Switzer, Catherine, *Unionists and the Great War Commemoration in the North of Ireland, 1914–1918* (Dublin: Irish Academic Press, 2007).

Tamari, Salim, *Year of the Locust: A Soldier's Diary and the Erasure of Palestine's Ottoman Past* (Berkeley and Los Angeles: University of California Press, 2011).

Thomson, Alistair, *Anzac Memories: Living with the Legend* (Melbourne: Oxford University Press, 1994).

Travers, Tim, *Gallipoli 1915* (Stroud: Tempus Publishing, 2001).

Travers, Tim, 'Liman Von Sanders, the Capture of Lieutenant Palmer, and Ottoman Anticipation of the Allied Landings at Gallipoli on 25 April 1915', *Journal of Military History* 65 (2001), 965–79.

Travers, Tim, *The Killing Ground: The British Army, the Western Front and the Emergence of Modern Warfare 1900–1918* (Barnsley: Pen & Sword Military Classics, 1987; repr. 2003).

Türkmen-Dervişoğlu, Gülay, 'Coming to Terms with a Difficult Past: The Trauma of the Assassination of Hrant Dink and Its Repercussions on Turkish National Identity', *Nations and Nationalism* 19/4 (2013), 674–92.

Üngör, Uğur Ümit, *The Making of Modern Turkey: Nation and State in Eastern Anatolia, 1913–1950* (Oxford: Oxford University Press, 2011).

Uyar, Mesut, 'Ottoman Arab Officers between Nationalism and Loyalty During the First World War', *War in History* 20/4 (2013), 526–44.

Uyar, Mesut, and Erickson, Edward J., *A Military History of the Ottomans: From Osman to Atatürk* (Santa Barbara, CA: Praeger Security International, 2009).

Waite, Major Fred, *The New Zealanders at Gallipoli* (Auckland: Whitcombe and Tombs Ltd, 1921).

Ward, Stuart, *Australia and the British Embrace: The Demise of the Imperial Ideal* (Melbourne: Melbourne University Press, 2001).

Ward, Stuart, 'Parallel Lives, Poles Apart: Commemorating Gallipoli in Ireland and Australia', in John Horne and Edward Madigan (eds), *Towards Commemoration: Ireland in War and Revolution, 1912–1923* (Dublin: Royal Irish Academy, 2013), 29–37.

West, Brad, 'Enchanting Pasts: The Role of International Civil Religious Pilgrimage in Reimagining National Collective Memory', *Sociological Theory* 26/3 (2008), 258–70.

West, Brad, 'Dialogical Memorialization, International Travel and the Public Sphere: A Cultural Sociology of Commemoration and Tourism at the First World War Gallipoli Battlefields', *Tourist Studies* 10/3 (2010), 209–25.

White, Jenny, *Muslim Nationalism and the New Turks* (Princeton: Princeton University Press, 2013).

Winter, J. M., *Sites of Memory, Sites of Mourning: The Great War in European Cultural History* (Cambridge: Cambridge University Press, 1995).

Winter, Jay, *The Great War and the British People* (London: Macmillan, 1986).

Winter, Jay, and Prost, Antoine, *The Great War in History: Debates and Controversies, 1914 to the Present* (Cambridge: Cambridge University Press, 2005).

Worthy, Scott, 'A Debt of Honour: New Zealanders' First Anzac Days', *New Zealand Journal of History* 36/2 (2002), 185–202.

Yılmaz, Hakan, *Secularism and Muslim Democracy in Turkey* (Cambridge: Cambridge University Press, 2009).

Ziino, Bart, 'Who Owns Gallipoli? Australia's Gallipoli Anxieties 1915–2005', *Journal of Australian Studies* 30/88 (2006), 1–12.

Ziino, Bart, *A Distant Grief: Australians, War Graves and the Great War* (Crawley, WA: University of Western Australia Press, 2007).

Zürcher, Erik, *Turkey: A Modern History* (London: I. B. Tauris, 1993; 2004 edn.).

Zürcher, Erik J., *The Young Turk Legacy and Nation Building: From the Ottoman Empire to Ataturk's Turkey* (London: I. B. Tauris, 2010).

PICTURE ACKNOWLEDGEMENTS

INDEX

Goldstein, Rabbi 108
Goliath, HMS 49
Goring, Percy 152
Gouraud, General Henri 49, 66
Greek inhabitants of Ottoman Empire 7,
    25, 27, 156, 159, 163, 170, 182, 185, 197
    (n. 15)
Gülcemal 162, 165
Güneş, Cengiz 181

Hamilton, General Sir Ian 72, 106, 129,
    140, 144
    appointment and journey to theatre
        17, 28
    Dardanelles Commission 72, 138
    decision to postpone attack, 18
        March 22, 28
    decisions, May and June 47, 48
    despatches 69, 144
    difficulties and errors 44, 50, 58, 66–7
    Gallipoli Diary (1920) 138–9
    interventions, 25 April 34, 41, 44
    personality and career 26, 66, 133
    plans amphibious attack 28–9, 35, 37
    sacked 64, 69
    Suvla 58, 61
Hammersley, Major General
    Frederick 60–1
Hart, Private Leonard 56–7
Haughey, Reverend Charles 147
Hawke, Bob 94–5, 99, 122, 123
Herbert, Aubrey 2
Herbert, A. P., *The Secret Battle* (1917) 135
heroism 3, 5, 40, 41, 45, 86, 91, 111, 134, 148
    heroic-romantic myth of Gallipoli 134,
        136, 138–40, 145
    pre Second World War references to
        heroism 71, 75, 82, 85, 108, 131, 142,
        143, 145, 155, 158, 161, 165, 166
    post Second World War references to
        heroism 93, 149, 167, 171, 173
    post 1990 references to heroism 124,
        178, 180
Hill, Reverend John 131, 141
Howard, John 94, 95–6, 99, 100, 123
Hughes, William (Billy) 82, 129
Hunter-Weston, Major General Sir
    Aylmer 24, 29, 34, 44

imperialism and commemoration 190–1
    post 1990 references (or absence
        thereof) 123, 125

post Second World War references (or
    absence thereof) 92, 100, 122, 147
pre Second World War references (or
    absence thereof) 81, 85, 114,
    117, 142
wartime references to empire during
    commemoration 76, 80, 108–9,
    129–30, 132, 136, 141
Imperial War Graves Commission 94, 163
Implacable, HMS 36, 43
India 7, 15, 67, 123, 133, 190
Inflexible, HMS 19, 21–2
Inglis, Ken 88
Inönü, İsmet 167
Ireland:
    'Celtic Tiger' period 151
    Easter Rising 108, 128, 144
    Home Rule 128
    Irish Free State 127, 128, 144
    War of Independence 5, 144
Irish Republican Army (IRA) 126, 128
Irresistible, HMS 19, 21–2

Jacka VC, Lance Corporal Albert 90–1
Jeffery, Major General Michael 99
Johnston, Brigadier General Francis 54
Jonquil HMS 58
Jones, Darcy 152
Joyce, Mr C. W. 84

Kannengiesser, Colonel Hans 21, 55–6
Kaya, Şükrü 165
Keating, Paul 95–6, 100, 123
Kemal, Lieutenant Colonel Mustafa (later,
    Atatürk) 5, 66, 140, 160–1, 166, 167,
    173–4, 179, 186, 189, 191
    commander, 19th Division 25, 26,
        42, 46
    decisive action 25 April 32–3, 35, 44
    lionized as Atatürk 161–2, 166
    modernizing reforms 159–60
    prominence in memory 155–6, 157,
        159, 162
    role at Chunuk Bair 8–10 August
        56–7, 121
    role at Suvla 60–1
    speech, 'Those heroes who shed their
        blood', 93–4, 95, 124, 165–6, 175, 176,
        186, 189
Keyes, Commodore Roger 17, 34, 43,
    64, 145
Kılıç, Suat 180